AF341612

Polyrotaxane and Slide-Ring Materials

Monographs in Supramolecular Chemistry

Series Editors:
Professor Philip Gale, *University of Southampton, UK*
Professor Jonathan Steed, *Durham University, UK*

Titles in this Series:

How to obtain future titles on publication:
A standing order plan is available for this series. A standing order will bring
delivery of each new volume immediately on publication.

For further information please contact:
Book Sales Department, Royal Society of Chemistry, Thomas Graham House,
Science Park, Milton Road, Cambridge, CB4 0WF, UK
Telephone: +44 (0)1223 420066, Fax: +44 (0)1223 420247
Email: booksales@rsc.org
Visit our website at http://www.rsc.org/Shop/Books/

Polyrotaxane and Slide-Ring Materials

By

Kohzo Ito
The University of Tokyo, Japan
Email: kohzo@k.u-tokyo.ac.jp

Kazuaki Kato
The University of Tokyo, Japan
Email: kato@molle.k.u-tokyo.ac.jp

Koichi Mayumi
The University of Tokyo, Japan
Email: kmayumi@molle.k.u-tokyo.ac.jp

Monographs in Supramolecular Chemistry No. 15

Print ISBN: 978-1-84973-933-7
PDF eISBN: 978-1-78262-228-4
ISSN: 1368-8642

A catalogue record for this book is available from the British Library

Published by The Royal Society of Chemistry,
Thomas Graham House, Science Park, Milton Road,
Cambridge CB4 0WF, UK

Registered Charity Number 207890

Visit our website at www.rsc.org/books

Printed in the United Kingdom by CPI Group (UK) Ltd, Croydon, CR0 4YY, UK

Preface

Supramolecular architectures with topological characteristics, such as catenanes and rotaxanes, have attracted considerable interest for a long time. They consist of connections with ring molecules rather than through conventional chemical bonds, whereby, as a consequence of topology, the molecular components are linked mechanically.

Polyrotaxane has a lot of rings threaded onto a backbone string creating a necklace-like topological supramolecule. Since these rings can slide along and rotate around the backbone axis freely, it has a huge amount of internal degree of freedom compared to usual polymers. This unique structure is maintained even in the slide-ring materials, a supramolecular network that is synthesized by cross-linking cyclic molecules in different polyrotaxanes. As a result, the slide-ring materials show quite different mechanical properties from conventional cross-linked polymeric materials such as gels, rubbers, resins, and so on.

This book is the first monograph of the slide-ring materials. Many books and reviews have been published on catenanes and rotaxanes, and some on polyrotaxanes. While they mainly described the synthesis and functionalization of them, here we focus on the physical properties of polyrotaxane and slide-ring materials, as well as the chemical ones. The peculiar topological architecture of polyrotaxane and slide-ring materials leads to a wide variety of unique physical properties, including mechanical, electrical, optical ones that are quite different from conventional polymeric materials. As a result, the slide-ring materials have been applied in coatings, adhesives, abrasives and actuators, and also for biomedical uses. Accordingly, I believe that this monograph will be helpful for industrial engineers developing novel polymeric materials to understand the physical properties and to know how to use polyrotaxane and slide-ring materials for their applications, as well as for

Monographs in Supramolecular Chemistry No. 15
Polyrotaxane and Slide-Ring Materials
By Kohzo Ito, Kazuaki Kato and Koichi Mayumi
© Kohzo Ito, Kazuaki Kato and Koichi Mayumi 2016
Published by the Royal Society of Chemistry, www.rsc.org

academic researchers, including students, interested in the supramolecular chemistry and/or polymer physics.

Three authors, including myself, participated in completing this monograph. Dr Kazuaki Kato mainly wrote Chapter 8 and a part of Chapter 4; Dr Koichi Mayumi chiefly took charge of Chapter 3 and a part of Chapter 4, and most of the rest was handled by myself. On behalf of the three, I would like to deeply acknowledge Dr Leanne Marle, who recommended me to write this monograph, and Dr Mina Roussenova for the continuous encouragement. Special thanks are also due to all staff and many students in our lab, who have yielded a lot of results based on this monograph. Finally, I wish to express my sincere gratitude and appreciation to Prof. Akira Harada, because he guided me to this fascinating research area. I started to investigate polyrotaxane in September 1995 immediately after I met Prof. Harada accidentally in an international symposium. This encounter resulted in the invention of the slide-ring materials.

Kohzo Ito
Tokyo

Contents

Monographs in Supramolecular Chemistry No. 15
Polyrotaxane and Slide-Ring Materials
By Kohzo Ito, Kazuaki Kato and Koichi Mayumi
© Kohzo Ito, Kazuaki Kato and Koichi Mayumi 2016
Published by the Royal Society of Chemistry, www.rsc.org

Introduction

Rotaxane and polyrotaxane have been investigated most extensively as typical mechanically interlocked molecular systems. They consist of a backbone string, rings, and two bulky end groups. The rings are threaded onto the axis string and are prevented from unthreading by the bulky end groups, as shown in Figure 1.1. The term "rotaxane" originates from the combination of the Latin roots "rota" and "axis", which mean "wheel" and "axle", respectively. When many rings are threaded onto an axis string, it is called a polyrotaxane, and appears as a necklace-like or rosary-like topological supramolecule. If there are no bulky end moieties, it is referred to as a pseudopolyrotaxane. When polyrotaxanes are cross-linked, the resulting supramolecular network is named as a slide-ring material.

The first report of supramolecules that have "chemical topology" was a catenane consisting of two (or more) interlocking macrocycles by Wasserman in 1960,[1] followed by the first concept of a rotaxane, which was proposed in 1961.[2] Rotaxane,[3] which was first named "hooplane", was synthesized by coupling the decane derivative and trityl chloride end groups with 2-hydroxycyclotriacontanone stabilized on a peptide resin. Since the interlocking relied on probability, the reaction resulted in a low yield of 6% even after 70 treatments. The term "rotaxane" was first used in another paper a while after that.[4]

The low yield of polyrotaxane synthesis was improved dramatically by the introduction of host–guest chemistry after the 1970s. A wide variety of cyclic molecules, such as cyclodextrins (cyclic oligosaccharides of D-glucose),[5] crown ethers[6] and cucurbiturils,[7] were found to form the inclusion complex formation with guest molecules. They include other substances spontaneously within the infinitesimal cavity by self-assembly, which is promoted by various strong interactions between host and guest moieties, overcoming the entropy loss. The polyrotaxane and slide-ring materials using cyclodextrins are especially highlighted in this monograph. Three kinds of

Monographs in Supramolecular Chemistry No. 15
Polyrotaxane and Slide-Ring Materials
By Kohzo Ito, Kazuaki Kato and Koichi Mayumi
© Kohzo Ito, Kazuaki Kato and Koichi Mayumi 2016
Published by the Royal Society of Chemistry, www.rsc.org

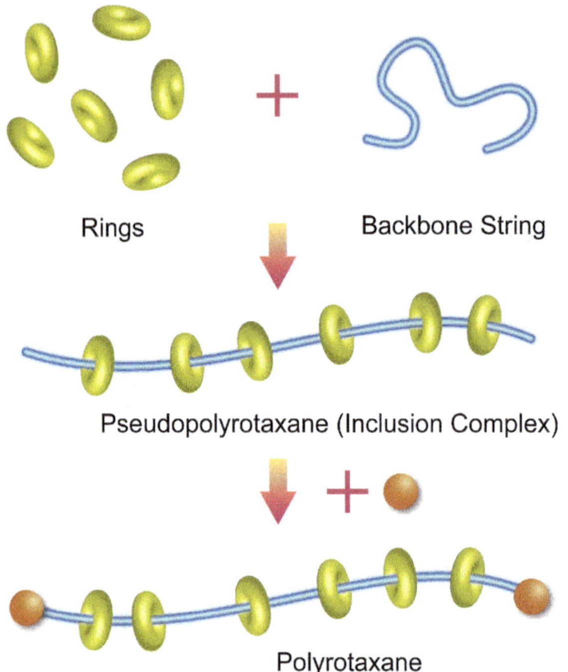

Figure 1.1 Schematic diagram of a synthesis scheme of polyrotaxane.

Table 1.1 Size dimensions of various cyclodextrins.[8]

CD type	Number of anhydroglucose units	Inner diameter of cavity, nm	Depth of cavity, nm
α	6	0.45	0.67
β	7	~ 0.70	~ 0.70
γ	8	~ 0.85	~ 0.70

cyclodextrins (CD) are commercially available: α-, β-, and γ-cyclodextrins for 6, 7, and 8 anhydroglucose units, respectively. Table 1.1 summarizes the dimensions of the three cyclodextrins. These cyclodextrins form the crystal systems characteristic of the cage, layer, and channel structures.[5,8] The most important feature of these cyclodextrins is their amphiphilic properties as they have a hydrophobic inside and a hydrophilic outside because of all the hydroxyl groups that are located outward of the cyclodextrin. Therefore, cyclodextrins are highly soluble in water but are not stable as long as the inside is occupied with water molecules. As a result, cyclodextrins include other guest molecules in their hydrophobic cavity, thereby leading to inclusion complex formation. This means that a wide variety of hydrophobic, low molecular compounds can be dispersed stably in water due to inclusion complex formation with cyclodextrins. Inclusion complex formation enabled Ogino *et al.* to first synthesize rotaxane using α- or β-cyclodextrins, and to achieve a relatively high yield.[9,10] The successful synthesis was followed by several researchers who reported rotaxanes of various backbone strings and cyclodextrins.

In 1990, Harada and Kamachi found that many α-cyclodextrins were threaded onto poly(ethylene glycol) to form polypseudorotaxanes.[11] The addition of an aqueous solution of poly(ethylene glycol) to a saturated aqueous solution of α-cyclodextrins yielded a white precipitate at room temperature, which was found to be pseudopolyrotaxane: many α-cyclodextrins were threaded onto a backbone poly(ethylene glycol). X-ray diffraction and ¹H NMR definitely proved the inclusion complex formation of α-cyclodextrins and poly(ethylene glycol), which was also supported by experimental results as precipitation was not observed following the addition of an aqueous solution of poly(ethylene glycol) modified with excessively large end groups such as 2,4-dinitrophenyl or 3,5-dinitrobenzoyl. Soon afterwards Harada *et al.* succeeded in the synthesis of polyrotaxane in a high yield by capping both ends of the pseudopolyrotaxane with dinitrobenzene in 1992.[12] It was a breakthrough for the application of polyrotaxane, as well as a fundamental study for the synthesis of polyrotaxane in high yield. It enabled us to apply the polyrotaxane structure to polymeric materials such as gels and elastomers. Around the same time, Wenz *et al.* independently synthesized other kinds of pseudopolyrotaxane and polyrotaxane with cyclodextrins.[13,14]

Since these findings, numerous investigations on the different combination of backbone polymers and cyclic molecules, especially cyclodextrins, have reported a wide variety of pseudopolyrotaxanes and polyrotaxanes. The common feature of the inclusion complex formation with cyclodextrins is the appearance of a water-insoluble white crystalline precipitate. It is remarkable that the synthesized polyrotaxanes are not soluble in water, although cyclodextrins and poly(ethylene glycol) in polyrotaxane are both highly soluble in water. Most polyrotaxanes dissolve only in dimethyl sulfoxide or sodium hydroxide aqueous solution. This suggests that the intramolecular aggregation of cyclodextrins by the strong hydrogen bonds between adjacent cyclodextrins results in a loss of the hydrophilic properties of the polyrotaxane due to the many hydroxyl groups on the cyclodextrins. The high yield of polyrotaxane is ascribed to the aggregation, which is based on the strong intramolecular interaction of the hydrogen bond, overcoming the entropy loss that accompanies inclusion complex formation. The channel-like structures of cyclodextrins further aggregate to form higher-order hexagonally packed columnar domains *via* the hydrophobic interactions and/or hydrogen bonds, which results in the white precipitate in water.

Another important feature of inclusion complex formation is the selectivity based on the host–guest chemistry between backbone polymers and cyclic moieties. For instance, poly(ethylene glycol) yields pseudopolyrotaxanes with α-cyclodextrin, but not with β-cyclodextrin, while poly(propylene glycol) forms only the inclusion complex with β-cyclodextrin.[15,16] This means that the optimal size fitting between the outside diameter of the backbone strings and the inside diameter of the rings is quite important in inclusion complex formation, which also supports the threading of cyclodextrins onto various backbone polymers.

As mentioned previously, the cyclodextrin-based inclusion complex formation is assisted by the strong hydrogen bonds between adjacent cyclodextrins. This tends to yield a polyrotaxane of the full inclusion or filling ratio, where the backbone polymer is covered almost fully with cyclodextrins. The filling ratio of polyrotaxane is strongly dependent on the backbone polymer species, the length of the backbone string, the end groups of the string, the kind of cyclodextrins and concentrations of them, and the solvent and temperature during inclusion complex formation. Regarding the polyrotaxane of poly(ethylene glycol) and α-cyclodextrin, backbone strings shorter than *ca.* 3000 in molecular weight produce almost fully covered polyrotaxane, *i.e.*, "dense polyrotaxane". However, as the length of poly(ethylene glycol) exceeds *ca.* 10 000 in molecular weight, the filling ratio decreases down to *ca.* 30% drastically, even if the saturated aqueous solution of α-cyclodextrin is used for the inclusion complex formation. This may arise from the following kinetics of the inclusion complex formation and the hydrophobicity of pseudopolyrotaxane. The aggregation among pseudopolyrotaxanes, resulting in the white precipitation, competes with threading of α-cyclodextrins onto poly(ethylene glycol). Since longer poly(ethylene glycol) needs longer to be fully covered with α-cyclodextrins, inclusion complex formation is disturbed in the middle by the aggregation among pseudopolyrotaxanes. Consequently, poly(ethylene glycol) longer than *ca.* 10 000 in molecular weight forms a polyrotaxane with a lower filling ratio. Interestingly, the high yield of polyrotaxanes with longer backbone polymers is achieved only at the filling ratio of 20 to 30%, almost independently of the length of the backbone poly(ethylene glycol), although the mechanism has not yet been elucidated.

As mentioned previously, we usually obtain a polyrotaxane with α-cyclodextrins sparsely distributed (*i.e.*, sparse polyrotaxane) when longer poly(ethylene glycol) is used as an axis polymer. In such a sparse polyrotaxane, α-cyclodextrins can slide along and rotate around the backbone poly(ethylene glycol) much more easily than in the dense polyrotaxane. In addition, another important feature of the polyrotaxane of poly(ethylene glycol) and α-cyclodextrins is that all reactive sites of the hydroxyl groups are located only on the α-cyclodextrins, not on poly(ethylene glycol), because both the reactive ends of poly(ethylene glycol) are capped with bulky end groups in the polyrotaxane. Okumura and Ito added cyanuric chloride as a cross-linker to a sodium hydroxide aqueous solution of polyrotaxane to form a polyrotaxane gel, which was a topological or slide-ring gel with figure-of-eight cross-links.[17] As schematically shown in Figure 1.2, the polymer chains in the topological gel are neither covalently cross-linked like chemical gels nor have attractive interactions with one another like physical gels. Instead, the backbone strings are mechanically or topologically interlocked by figure-of-eight movable cross-links. The cross-links can pass along the backbone strings freely to equalize the tension of the threaded backbone strings just

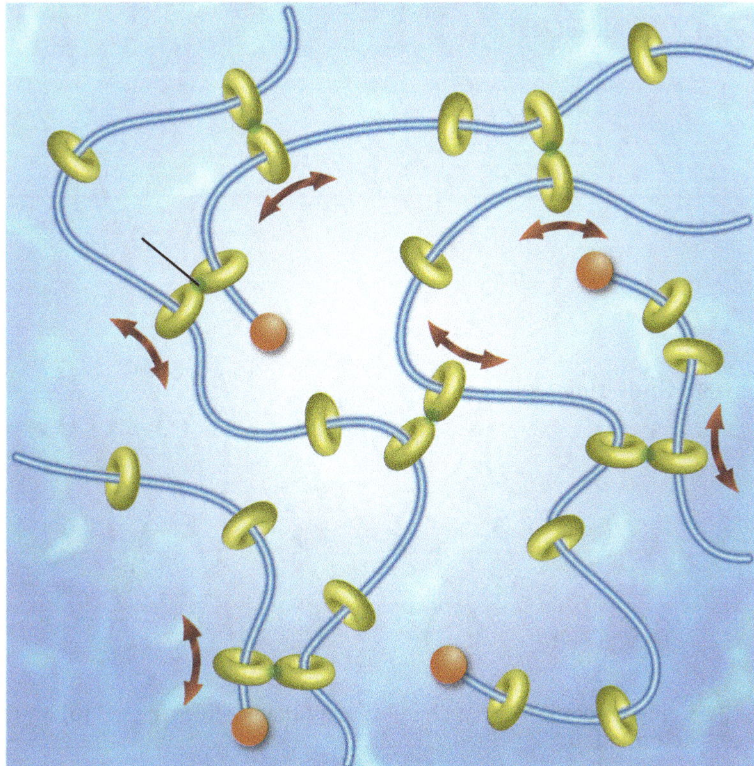

Figure 1.2 Schematic diagram of the topological or slide-ring gel.

like pulleys. As a result, the nanosized heterogeneity in structure and stress may be automatically homogenized in the topological gel.

Figure 1.3 schematically compares the chemical and topological gels with regards to tensile deformation. In the chemical gel, the heterogeneous polymer length between fixed cross-links causes a concentration of tensile strength on shorter network strands, resulting in breakage. On the other hand, the network strands in the topological gel can pass through the movable cross-links cooperatively like pulleys, changing the strand length, which is called the pulley effect.[17] This enables network strands to equalize the tension not only in a single polymer chain, but also among adjacent strands interlocked by the movable cross-links. The unique molecular features of the topological gel yield various physical properties, especially mechanical ones.

The other important feature of the topological gel is the entropy of the rings, as described in more detail in Chapter 4. Polyrotaxane intrinsically has two kinds of entropy: the conformational entropy of the backbone string, and the distribution or arrangement entropy of the rings. These two entropies are almost independent of each other in the polyrotaxane. However, the movable cross-links in the topological gel yield strong coupling between

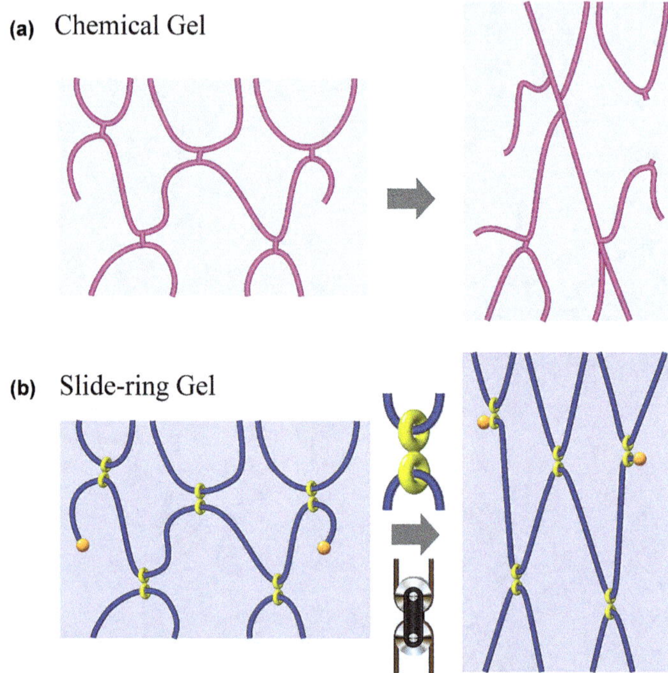

Figure 1.3   Comparison between a chemical gel (a) and slide-ring gel (b) on tensile deformation.

these two kinds of entropies. It is to be noted that all α-cyclodextrins are not cross-linked and there are many uncross-linked free rings in the topological gel network. The pulley effect can change the length of the network strands but cannot change the number of free rings between cross-links. As a result, the pulley effect competes with the entropy of the rings, which yields a wide variety of peculiar mechanical properties such as the sliding transition and siding elasticity.

The fundamental concept of the topological gel can be extended to polymeric materials in the absence of solvent. In particular, when it is applied to commercially available items, we do not have to use polyrotaxane in isolation. Instead, the concept can be incorporated into usual polymeric materials, such as conventional elastomers and resins, by cross-linking polyrotaxane with them, as schematically shown in Figure 1.4, which improves the mechanical properties drastically. These polymeric materials including polyrotaxane are referred to as slide-ring materials in a wider sense. However, as the commercial and industrial demand for slide-ring materials has increased, the mass production and cost reduction of polyrotaxane have become crucial issues. Araki *et al.* developed an efficient synthesis scheme of polyrotaxane in an extremely high yield of over 95%,[18] and polyrotaxane grafted with origo-ε-caprolactone has proved useful for solid-state slide-ring materials.[19] With Araki's scheme, some major chemical companies have

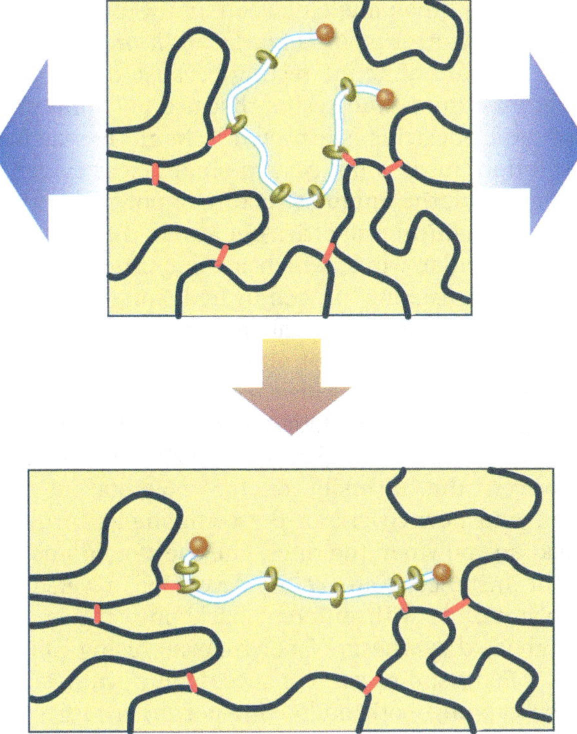

Figure 1.4 Cross-linking of polyrotaxane with conventional polymers. This results in the improvement of mechanical properties without changing the chemical properties and manufacture processes. Thus, the concept of slide-ring materials can be applied to a wide variety of polymeric materials at low cost.

successfully produced origocaprolactone-grafted polyrotaxane in the ton scale, which is commercially available at a reasonable price. As a result, slide-ring materials are now applied to a variety of items such as coatings, paints, adhesives, abrasives, rubbers, resins, soft actuators, and so on. For example, the scratch-resist properties of the self-restoring slide-ring coating were adopted in the upper coatings of mobile phones. In addition, the polyrotaxane of poly(ethylene glycol) and α-cyclodextrins is expected to have high biosafety and biocompatibility because all the components are approved for medical use and/or as food additives. Accordingly, the biomedical applications of polyrotaxane and slide-ring materials are now proceeding.

Concurrently with the creation and enhancement of these applications, the fundamental researches on polyrotaxane and slide-ring materials have been promoted extensively and in an increasing manner. As a result, various new and abnormal mechanical properties of the slide-ring materials have been found, including in the biaxial stretching test, the swelling behavior, and the permeation flow properties, while the peculiar structures of polyrotaxane and slide-ring materials have been investigated including the molecular dynamics by small angle neutron and X-ray scatterings, dynamic light scattering, and neutron spin echo. In addition, the solid-state properties of polyrotaxane are quite interesting because they contribute to various functions of elastomers and resins that use polyrotaxane. The sliding and rotation of α-cyclodextrins is expected to be no less active in the solid state than in solution but it is not easy to obtain clear evidence of the motion. On the other hand, the unique structure of polyrotaxane does not only contribute to the mechanical properties but also to the electrical and optical ones. Insulated molecular wires, which are identified as polyrotaxane or pseudopolyrotaxane consisting of conjugated conducting polymers and cyclic molecules, show unique electrical and optical properties, which could be applied to molecular electronic or optical devices in the future.

Most of these physical properties have been measured for the polyrotaxane of poly(ethylene glycol) and α-cyclodextrins partly because it can be synthesized much more easily in higher yield than those of the other polyrotaxanes. However, the diversity of the polyrotaxane synthesis has progressed lately and now covers all the components of polyrotaxane, including the backbone polymer, the rings, and the stoichiometry of these two components. For instance, Kato *et al.* developed a versatile scheme to synthesize a polyrotaxane with different backbone strings.[20] Ogoshi *et al.* achieved the high-yield synthesis of polyrotaxane using pillar[5]arene as a cyclic molecule.[21] Feng and coworkers successfully synthesized a variety of polyrotaxanes by capping both ends with polymers rather than with not bulky groups.[22] These polymers added various novel functions to the polyrotaxane. In addition, Kidowaki *et al.* modified polyrotaxane with liquid crystalline mesogens.[23] The diversity of polyrotaxane components enhances the findings of novel physical properties. In particular, the different combinations of the backbone string and rings are expected to change the mobility of the rings, which may be controlled by external stimuli. More degrees of freedom and peculiar molecular motions in the components give polyrotaxanes with crucial differences from usual polymers.

As stated previously, polyrotaxane and slide-ring materials are establishing a new field in the inter-disciplinary area between supramolecular chemistry and polymer physics. In this monograph, we will introduce the cutting-edge trends of this new emerging field, especially focusing on the physical and material aspects, and on applications of polyrotaxane and slide-ring materials, together with the chemical view, although it is still in the middle of establishment. Regarding the chemistry of polyrotaxane, such as its synthesis and functionalization, several excellent books and many reviews have been published to date.[24–38] In Chapter 2, we introduce the thermodynamics of inclusion complex formation between a backbone string and many rings, and between a backbone string and a molecular tube, followed by a sliding transition in a polyrotaxane having the block copolymer as a backbone string. Chapter 3 deals with the structural analysis and dynamics of polyrotaxane and slide-ring materials by scattering experiments with lasers, synchrotron radiation, and neutrons. Neutron scattering is a powerful tool, especially for multi-component system such as polyrotaxane and slide-ring materials. The mechanical properties of the slide-ring materials are described in detail in Chapter 4. The pulley effect and entropy of the rings yields a wide variety of peculiar mechanical properties, which have not been observed in usual polymeric materials. In Chapter 5, the solid-state properties of polyrotaxane are summarized, which suggest the active sliding or rotational motion of rings and/or the backbone string in the polyrotaxane. In Chapter 6, we introduce the structure and electrical and/or optical properties of insulated molecular wires. The high performance of the insulated molecular wire exhibits a potential application of polyrotaxane and pseudopolyrotaxane to molecular electronic and optical devices in the future. Chapter 7 focuses on the chemistry of polyrotaxane and slide-ring materials, where the diversity in the backbone string, the rings, and the stoichiometry of these two components is expanded increasingly to control the molecular motion. Finally, we show the present and future applications of polyrotaxane and slide-ring materials in Chapter 8. The polyrotaxane and slide-ring gel consisting of poly(ethylene glycol) and α-cyclodextrins can contribute to the biomedical applications in the future because of their high biosafety and biocompatibility, and slide-ring materials are already commercially available and applied to a wide variety of items.

Since Goodyear discovered the covalent cross-links of natural rubber with sulfur in 1839, cross-links have been considered fixed on polymer chains in polymeric network materials. The new concept of movable cross-links appeared on the basis of polyrotaxane with a topological supramolecular architecture in *ca.* 2000. It has been recently reported that polymeric materials incorporating the polyrotaxane structure as the movable cross-linker improve with regards to a wide variety of mechanical properties such as strain at break and fracture toughness. This may be because the movable cross-links relax the heterogeneous stress in the polymeric materials. Accordingly, the movable cross-link of slide-ring materials, a brand-new concept in polymer science and technology, may bring a breakthrough to solve the problem of cross-linking heterogeneity inherent in chemical cross-linking since Goodyear's discovery.

References

1. E. Wasserman, *J. Am. Chem. Soc.*, 1960, **82**, 4433.
2. H. L. Frisch and E. Wasserman, *J. Am. Chem. Soc.*, 1961, **83**, 3789.
3. T. Harrison and S. Harrison, *J. Am. Chem. Soc.*, 1967, **89**, 5723.
4. G. Schill and H. Zollenkopf, *Liebigs Ann. Chem.*, 1969, **721**, 53.
5. M. L. Bender and M. Komiyama, *Cyclodextrin Chemistry*, Springer-Verlag, Berlin, 1978; J. Szejtli, *Chem. Rev.*, 1998, **98**, 1743.
6. P. R. Ashton, E. J. T. Chrystal, P. T. Glink, S. Menzer, C. Schiavo, N. Spencer, J. F. Stoddart, P. A. Tasker, A. J. P. White and D. J. Williams, *Chem.-Eur. J.*, 1996, **2**, 709.

7. J. Lagona, P. Mukhopadhyay, S. Chakrabarti and L. Issacs, *Angew. Chem., Int. Ed.*, 2005, **44**, 4844.

8. G. Wenz, B.-H. Han and A. Müller, *Chem. Rev.*, 2006, **106**, 782.

9. H. Ogino, *J. Am. Chem. Soc.*, 1981, **103**, 1303.

10. H. Ogino, *New J. Chem.*, 1993, **17**, 683.

11. A. Harada and M. Kamachi, *Macromolecules*, 1990, **23**, 2821.

12. A. Harada, J. Li and M. Kamachi, *Nature*, 1992, **356**, 325.

13. G. Wenz and B. Keller, *Angew. Chem., Int. Ed.*, 1992, **31**, 197.

14. G. Wenz and B. Keller, *Macromol. Symp.*, 1994, **87**, 11.

15. A. Harada, *Coord. Chem. Rev.*, 1996, **148**, 115.

16. A. Harada, *Adv. Polym. Sci.*, 2006, **201**, 1.

17. Y. Okumura and K. Ito, *Adv. Mater.*, 2001, **13**, 485.

18. J. Araki, C. Zhao and K. Ito, *Macromolecules*, 2005, **38**, 7524.

19. J. Araki, R. Kataoka and K. Ito, *Soft Matter*, 2008, **4**, 245.

20. K. Kato, H. Komatsu and K. Ito, *Macromolecules*, 2010, **43**, 8799.

21. T. Ogoshi, Y. Nishida, T. Yamagishi and Y. Nakamoto, *Macromolecules*, 2010, **43**, 7068.

22. X. Tong, X. Zhang, L. Ye, A. Zhang and Z. Feng, *Polymer*, 2008, **49**, 4489.

23. M. Kidowaki, T. Nakajima, J. Araki, A. Inomata, H. Ishibashi and K. Ito, *Macromolecules*, 2007, **40**, 6859.

24. J.-M. Lehn, *Supramolecular Chemistry: Concepts and Perspectives*, VCH, Weinheim, 1995.

25. *Molecular Catenanes, Rotaxanes and Knots*, ed. J.-P. Sauvage, C. Dietrich-Buchecker, Wiley, Chichester, 1999.

26. H. W. Gibson and H. Marand, *Adv. Mater.*, 1993, **5**, 11.

27. H. W. Gibson, M. C. Bheda and P. T. Engen, *Prog. Polym. Sci.*, 1994, **19**, 843.

28. G. Wenz, *Angew. Chem., Int. Ed.*, 1994, **33**, 803.

29. A. Harada, *Coord. Chem. Rev.*, 1996, **148**, 115.

30. S. A. Nepogodiev and J. F. Stoddart, *Chem. Rev.*, 1998, **98**, 1959.

31. M. Raymo and J. F. Stoddart, *Chem. Rev.*, 1999, **99**, 1643.

32. T. Takata, N. Kihara and Y. Furusho, *Adv. Polym. Sci.*, 2004, **17**, 1.

33. F. Huang and H. W. Gibson, *Prog. Polym. Sci.*, 2005, **30**, 982.

34. T. Takata, *Polym. J.*, 2006, **38**, 1.

35. A. Harada, *Adv. Polym. Sci.*, 2006, **201**, 1.

36. H. Tian and Q.-C. Wang, *Chem. Soc. Rev.*, 2006, **35**, 361.

37. J. Araki and K. Ito, *Soft Matter*, 2008, **4**, 245.

38. A. Harada, Y. Takashima and H. Yamaguchi, *Chem. Soc. Rev.*, 2009, **38**, 875.

Thermodynamics of Inclusion Complex Formation and the Sliding Transition in Polyrotaxane

2.1 Inclusion Complex Formation of a Polymer and Rings

Inclusion complex formation between ring molecules and an axis polymer string is the key technology to make rotaxane or polyrotaxane. Actually, Harada and Kamachi found that α-cyclodextrin (α-CD) and poly(ethylene glycol) (PEG) formed pseudopolyrotaxane by inclusion complex formation in water in 1990,[1] which led to a great number of researches on pseudopolyrotaxanes of various ring molecules and linear polymer backbones. Therefore, it is important to investigate the thermodynamics of inclusion complex formation. The degree of freedom of ring molecules is changed drastically before and after the inclusion complex formation of polyrotaxane. Ring molecules threaded together with a polymer string can move only one-dimensionally along the backbone polymer after inclusion complex formation. This means that the inclusion complex formation is enthalpically favorable but entropically unfavorable. Consequently, the inclusion complex formation needs some attractive interactions between the ring and string molecules, and/or between rings.

The competition between enthalpy and entropy in the inclusion complex formation or in polyrotaxane should yield a variety of transitions. For example, ring molecules are likely to be threaded onto a backbone polymer at low temperature, and dissociated from it at high temperature. The

---

Monographs in Supramolecular Chemistry No. 15
Polyrotaxane and Slide-Ring Materials
By Kohzo Ito, Kazuaki Kato and Koichi Mayumi
© Kohzo Ito, Kazuaki Kato and Koichi Mayumi 2016
Published by the Royal Society of Chemistry, www.rsc.org

inclusion–dissociation behavior and the transition temperature are strongly dependent on the interaction energy between the ring and string molecules, and/or between adjacent rings. The stronger the attractive interaction between adjacent ring molecules, the more cooperative the inclusion–dissociation behavior becomes. In the strongest limit, rings can form a tube. Actually, since α-cyclodextrins have many hydroxyl groups at both sides of the ring, they can form several hydrogen bonds between adjacent α-cyclodextrins to join α-cyclodextrins together in a tubular structure on the polymer chain. Harada and Kamachi reported the synthesis of a molecular tube of α-cyclodextrins by cross-linking adjacent α-cyclodextrins in a polyrotaxane composed of α-cyclodextrins and poly(ethylene glycol) in 1993.[2] The molecular tube should show the inclusion complex formation is different from rings with a polymer string.

When the inside diameter of the rings is as small as the α-cyclodextrins, a polymer chain forming an inclusion complex with α-cyclodextrins has an extended conformation, such as a planar zig-zag one, with no degrees of freedom other than a translational motion along its longitudinal axis, which is reminiscent of the well-known "tube" model for the entanglement effect of polymer chains.[3] As the temperature increases, the polymer chain dissociates from the α-cyclodextrins and recovers its intrinsic entropy of a random conformation in solution. The entropy increases linearly with increasing polymer contour length if the chain has a random conformation. Accordingly, a long polymer chain yields a drastic change of entropy in the inclusion or dissociation process, which leads to the inclusion–dissociation transition.

Okumura *et al.* theoretically investigated the inclusion and dissociation behaviors in the inclusion complex formation of polymer strings and ring molecules in solution.[4] An ingenious theoretical model was proposed for calculation of the partition function of the system with many strings, including the different number of rings randomly distributed on them. To evaluate the free energy of a system consisting of polymer strings, rings, and solvent molecules, the theory adopted a lattice model assuming that the string is composed of segments with the same size as the solvent or ring molecule occupying one site on the lattice. For simplicity, it was assumed that the inside of a ring molecule is filled with solvent molecules or a polymer string segment, and that a string consisting of N segments occupies N connected lattice points, as schematically shown in Figure 2.1, where $\Delta\varepsilon_{in}$ is the inclusion energy per segment, ΔS is the conformational entropy per segment lost by the inclusion of a ring molecule, and $\Delta\varepsilon_a$ is the interaction energy between adjacent rings. Then, $\Delta F_{in} = \Delta\varepsilon_{in} - T\Delta S$ represents the inclusion free energy per segment ΔF_{in} at temperature T. Incidentally, the inclusion energy $\Delta\varepsilon_{in}$ is defined by $\Delta\varepsilon_{in} = \varepsilon_{pi} - \varepsilon_{si} + \varepsilon_{ss} - \varepsilon_{ps}$, where ε_{pi}, ε_{si}, ε_{ss} and ε_{ps} are, respectively, the interaction energies of the polymer string segment–ring inside (pi), the solvent molecule–ring inside (si), the solvent molecule–solvent molecule (ss), and the polymer string segment segment–solvent molecule (ps).

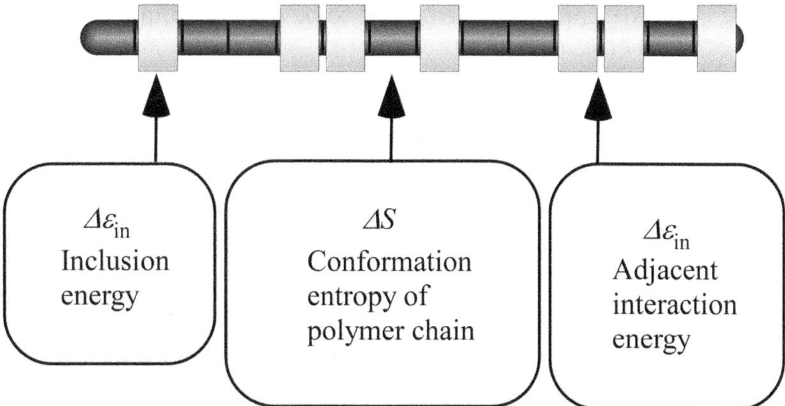

Figure 2.1 A lattice model for the inclusion complex of cyclodextrin molecules and a polymer chain.[4]
Reproduced by permission of John Wiley & Sons, Inc.

Next the theory gave us the free energy of a system containing N_{p} linear polymer strings and N_{r} ring molecules, where $N_{\mathrm{t}} = NN_{\mathrm{p}}$ is the total number of polymer string segments in the solution. Let Ω be the total number of lattice sites in the system and i the total number of segments included by rings. Then $f_{\mathrm{i}} = i/N_{\mathrm{t}}$ and $\alpha = N_{\mathrm{r}}/N_{\mathrm{t}}$ denote the number ratios of the segments included by rings to the total polymer string segments, *i.e.*, the filling ratio and the ratio of the total rings to the total polymer string segments in the whole system, and $\phi_p = N_{\mathrm{t}}/\Omega$ represents the volume fraction of polymer strings in solution. The partition function of inclusion complexes consisting of N_{p} polymer strings and i ring molecules including polymers was evaluated using the following procedure. Since the polymer chain has a finite length, it is difficult to strictly treat the adjacent interaction at both ends of the chain, *i.e.*, the end effect. To avoid it, all N_{p} polymer strings are connected into a huge hypothetical circular molecular necklace containing i ring molecules, and rings are aggregated into h groups by the adjacent interaction. Then, the partition function $R(i)$ of the circular molecular necklace is exactly given by:

$$R(i) = \sum_{h=1}^{i \text{ or } N_{\mathrm{t}}-i} \binom{i}{h}\left[N_{\mathrm{t}} - \binom{i}{h}\right] \frac{h}{i(N_{\mathrm{t}} - i)} \exp[i\Delta\bar{F}_{\mathrm{in}} + (i - h)\Delta\bar{\varepsilon}_{\mathrm{a}}], \qquad (2.1)$$

where $\Delta\bar{F}_{\mathrm{in}}$ and $\Delta\bar{\varepsilon}_{\mathrm{a}}$ are the reduced free and interaction energies given by $\Delta\bar{F}_{\mathrm{in}} = \Delta F_{\mathrm{in}}/k_{\mathrm{B}}T$ and $\Delta\bar{\varepsilon}_{\mathrm{a}} = \Delta\varepsilon_{\mathrm{a}}/k_{\mathrm{B}}T$, respectively, with the thermal energy $k_{\mathrm{B}}T$. The first two factors on the right-hand side represent the numbers of ways for cutting i included polymer segments and $N_{\mathrm{t}} - i$ free segments into h groups, as schematically shown in Figure 2.2. The next h indicates the number of ways for combining each h group into the huge circular necklace with N_{t} polymer segments and i ring molecules. The last factor represents the inclusion free energy and the adjacent interaction energy. When N_{t} and

i are quite large, $R(i)$ can be approximated to be the maximum value of R_h at the average number h_0 of h groups:

$$R(i) \approx \binom{i}{h_0}\left[N_t - \binom{i}{h_0}\right] \frac{h}{i(N_t - i)} \exp[i\Delta\bar{F}_{in} + (i - h_0)\Delta\bar{\varepsilon}_a], \qquad (2.2)$$

where h_0 is given from the condition $\partial R_h/\partial h = 0$ as:

$$h_0 = N_t\left[\frac{1 - \sqrt{1 - 4[1 - \exp(\Delta\bar{\varepsilon}_a)]f_i(1 - f_i)}}{2[1 - \exp(\Delta\bar{\varepsilon}_a)]}\right] \qquad (2.3)$$

Now we cut the hypothetical circular molecular necklace into N_p separate inclusion complexes, as shown in Figure 2.2. Since all polymer chains consist of N segments, there are N ways of cutting the circular necklace.

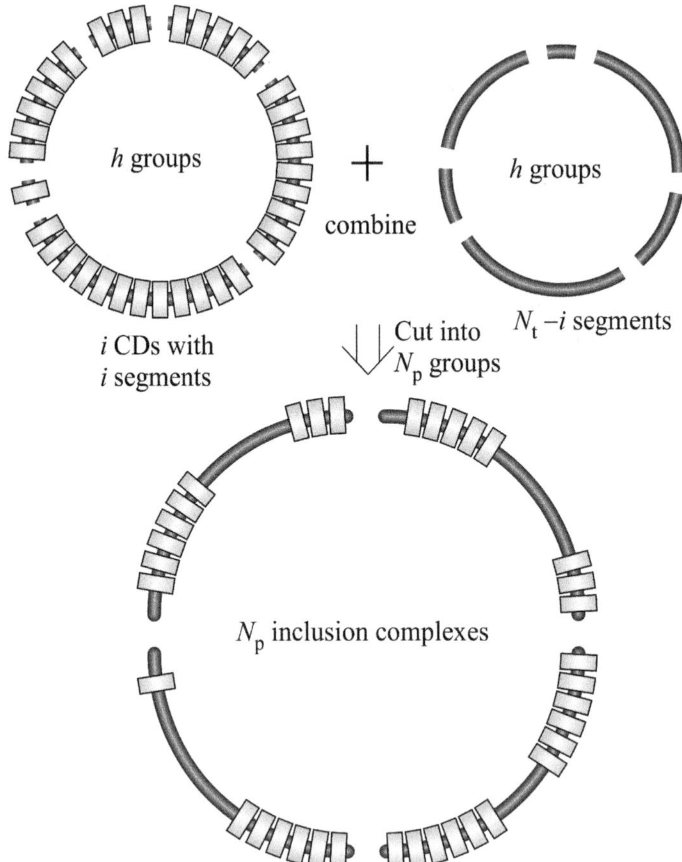

Figure 2.2   The procedure for calculating the partition function $N(i)C(i)$ of inclusion complexes. The two upper circles respectively represent the two factors on the right-hand side of eqn (2.1).[4]
Reproduced by permission of John Wiley & Sons, Inc.

Every time the adjacent ring molecules are cut off, the adjacent interaction energy $\Delta\varepsilon_a$ is lost; the probability of cutting the adjacent ring molecules is given by $f_i - f_0 = (i - h_0)/N_t$. Taking these factors into consideration, the partition function of the separation into N_p inclusion complexes is given by:

$$C(i) = N \exp\left[-N_p \frac{i - h_0}{N_t} \Delta\bar{\varepsilon}_a\right] \frac{\Omega^{N_p}}{N_p!}. \tag{2.4}$$

The last factor represents the number of ways of placing N_p inclusion complexes on the lattices.

Next, the theory focuses on $N_r - i$ ring molecules dissociated from the polymer strings. It was assumed that the adjacent interaction may connect some dissociated ring molecules to the tubular structures. Here, dissociated $N_r - i$ ring molecules are connected into d groups. Then, $N_r - i - d$ adjacent bonds are formed between the dissociated ring molecules. Thus, the partition function $D(i)$ of the dissociated ring molecules is given by:

$$D(i) = \sum_{d=0}^{N_r-i} \left[N_r - \binom{i}{d}\right] \frac{\Omega^d}{(N_r - i)d!} \exp[(N_r - i - d)\Delta\bar{\varepsilon}_a]. \tag{2.5}$$

Similar to $R(i)$, $D(i)$ is approximately reduced to the maximum value at the average number d_0 of d groups.

Then, the total partition function of the whole system is given as $Z(i) = R(i)C(i)D(i)$. The logarithm of the total partition function gives us the reduced (*i.e.*, dimensionless) free energy $\bar{F}(f_i) = F(f_i)/(N_t k_B T)$ per polymer string segment as a function of the filling ratio f_i:

$$\bar{F}(f_i) = -\left[f_i(1 - q) + (f_i - f_0)\left(1 - \frac{1}{N}\right)q + (\alpha - f_i - f_d\alpha)q\right]\Delta\bar{F}_{\text{sum}}$$

$$- f_i \ln f_i - (1 - f_i)\ln(1 - f_i) + (f_i - f_0)\ln(f_i - f_0) + 2f_0 \ln f_0 \tag{2.6}$$

$$+ (1 - f_i - f_0)\ln(1 - f_i - f_0) - f_d\alpha(\ln\phi_p - 1) - (\alpha - f_i)\ln(\alpha - f_i)$$

$$+ (\alpha - f_i - f_d\alpha)\ln(-f_i - f_d\alpha) + 2f_d\alpha \ln f_d\alpha,$$

where $f_d = d_0/N_r$, $\Delta\bar{F}_{\text{sum}} = \Delta\bar{F}_{\text{in}} + \Delta\bar{\varepsilon}_a$, and $q = \Delta\bar{\varepsilon}_a / \Delta\bar{F}_{\text{sum}}$ is the ratio of the interaction energy between adjacent rings to the inclusion free energy per segment.

Figure 2.3 shows the α dependence of f_i, obtained from the condition $\partial\bar{F}(f_i)/\partial f_i = 0$, for different values of q and a constant value of $\Delta\bar{F}_{\text{sum}}$. When $q = 0$, f_i increases in proportion to α at first, and is saturated gradually like usual inclusion complex formations between α-cyclodextrin and short polymer strings. This is because rings on a polymer string do not interact with each other in $q = 0$. As q rises, the inclusion–dissociation behavior becomes cooperative and sharp with increasing adjacent interactions, and f_i is saturated earlier. This indicates that the dissociated ring molecules are

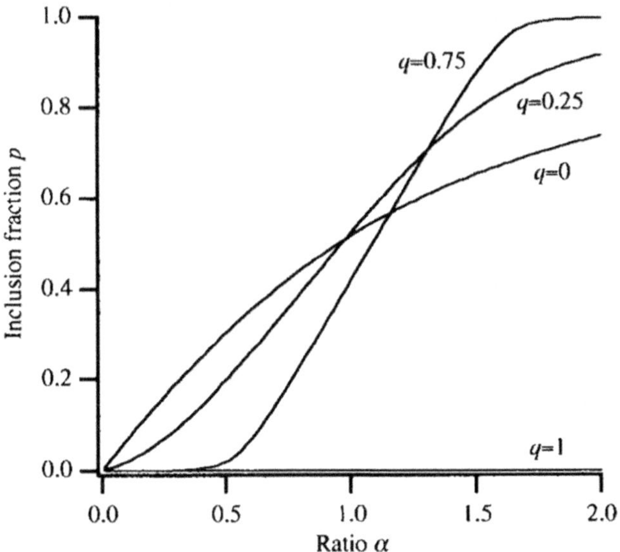

Figure 2.3   The α dependence of the inclusion fraction p for different values of q.[4] Reproduced by permission of John Wiley & Sons, Inc.

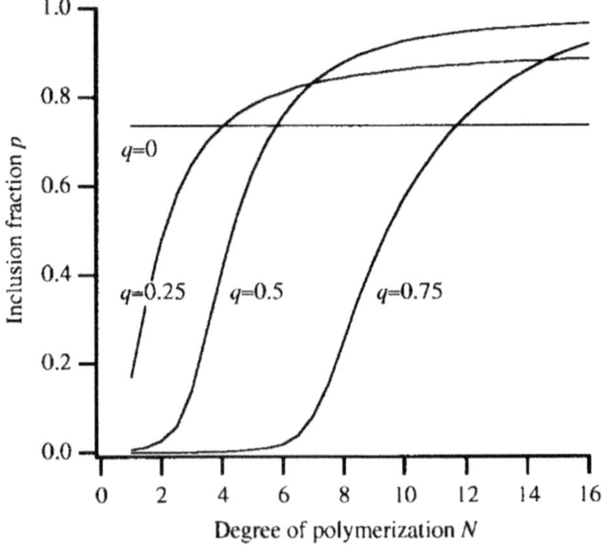

Figure 2.4   The N dependence of the inclusion fraction p for different values of q.[4] Reproduced by permission of John Wiley & Sons, Inc.

connected into the tubular structures by the adjacent interactions, and sharply includes the polymer chains as q increases.

Figure 2.4 shows the N dependence of f_i for different values of q and a constant value of $\Delta \bar{F}_{\text{sum}}$. When $q = 0$, f_i is completely independent of N.

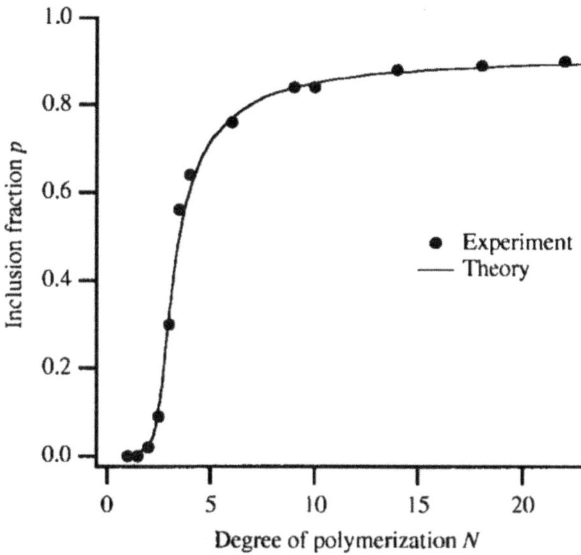

Figure 2.5 The experimental results of the N dependence of p by Harada *et al.*[5] and the best-fitted curve obtained from eqn (2.6).[4]
Reproduced by permission of John Wiley & Sons, Inc.

As q increases, a threshold of the inclusion complex formation appears at a value of N and shifts to larger N values. Figure 2.5 shows the experimental results of the N dependence of f_i in the inclusion complex formation between α-cyclodextrin and poly(ethylene glycol) by Harada *et al.*[5] and the best-fitted curve obtained from eqn (2.6). As shown in Figure 2.5, the theory indicated a good agreement with the experimental results. From this fitting procedure, the inclusion free energy $\Delta \bar{F}_{\mathrm{in}} = 3.24$ for an α-cyclodextrin molecule and the interaction energy $\Delta \bar{\varepsilon}_a = 5.76$ between adjacent cyclodextrins.

Next, Okumura *et al.* investigated the temperature dependence of the inclusion behavior.[4] Figure 2.6 shows the reduced temperature $\bar{T}$ dependence of f_i for different values of r and a constant value of ΔS, where $r = \Delta \bar{\varepsilon}_a / (\Delta \bar{\varepsilon}_{\mathrm{in}} + \Delta \bar{\varepsilon}_a)$ and $\bar{T} = T/T_0$ with the transition temperature $T_0 = (\Delta \bar{\varepsilon}_{\mathrm{in}} + \Delta \bar{\varepsilon}_a)/\Delta S$. When r is small, f_i decreases gradually like usual inclusion complex formations between α-cyclodextrin and short polymer strings. As r increases, f_i decreases cooperatively and sharply, and is decayed earlier. These tendencies are similar to the inclusion behavior between the molecular tube and the linear polymer chain as mentioned in the following pages.[6]

In summary, the inclusion–dissociation behavior in the complex formation between ring molecules and the polymer strings in solution was investigated theoretically by the lattice model. Without the interaction between adjacent rings, the rings are dissociated from or included into the polymer chain gradually as the amount of rings or temperature varied. However, the inclusion–dissociation behavior became cooperative and sharp with increasing adjacent interaction.

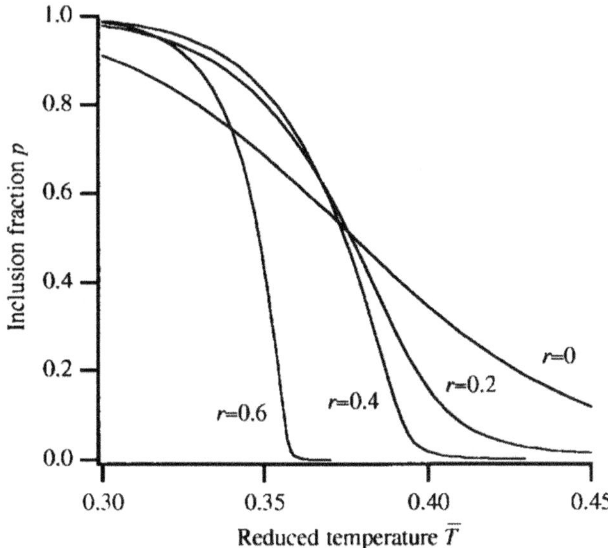

Figure 2.6   The reduced temperature $\bar{T}$ dependence of the inclusion fraction p for different values of r.[4]
Reproduced by permission of John Wiley & Sons, Inc.

2.2   Inclusion Complex Formation of Polymer and Tube

Nanotubes, which are fine capillaries with an inside diameter of the nanometer order, have attracted the great interest of many scientists because of their peculiar structures. A typical example is a carbon nanotube formed by an arc discharge method. Recently, a new series of nanotubes with diameters smaller than carbon ones have been chemically synthesized from cyclodextrins of cyclic form by Harada *et al.*[2] They prepared a polyrotaxane supramolecule in which cyclodextrins were threaded on a polymer chain with bulky ends and obtained a molecular tube by cross-linking the adjacent cyclodextrin units in the polyrotaxane. By removing the bulky ends of the polymer thread, the tube was unthreaded and acted as a host for reversible binding of small molecules. This molecular nanotube, which was soluble in several kinds of solvents such as water, had a constant inside diameter (*e.g.*, 0.45 nm for α-cyclodextrin) and a longitudinal length of the submicron order, which was controllable by varying the length of the polymer chain as a mold, and could form an inclusion complex with a linear polymer chain.

To evaluate the free energy of a system consisting of molecular nanotubes, polymer chains, and solvent molecules, we adopt a lattice model assuming that the tube and the chain are composed of segments with the same size as the solvent molecule occupying one site on the lattice, similar to the previous section. For simplicity, the tube is assumed to be fully rigid and has a longitudinal length equal to the contour length of the chain; namely, both the tube and the chain consist of N segments occupying N connected lattice points. The inside of the tube is filled with chain segments or solvent molecules, so that the tube segment also occupies one lattice site together with the chain segment or the solvent molecule. While the chain dissociated from the tube has a flexible or coiled conformation with many internal degrees of freedom, inclusion into the tube drastically changes the chain conformation to a rigid rod-like one. Then, the number of ways of placing the rod-like chain fully included into a tube is estimated to be Ωz, which is much smaller than $\Omega z(z-1)^{N-2}$ for a dissociated flexible chain, where Ω is the total number of lattice sites and z is the coordination number of the lattice. Let N_t and N_c be the total numbers of tubes and polymer chains, respectively, and i be the number of tubes forming inclusion complexes with polymer chains. Then, $p = i/N_t$ and $\alpha = N_c/N_t$ denote the number ratios of the inclusion tubes to the total ones and of the total polymer chains to the total tubes, respectively. By using Flory–Huggins theory, we estimate the p and α dependence of the free energy F of the system, in which parts of the polymer chains are dissociated from the tubes and interact with each other.

The inclusion energy difference ΔE_{in} of a chain into a tube from the full dissociation state $(i = 0)$ is given by

$$\Delta E_{in} = -(z - 2)(\epsilon_{ci} - \epsilon_{si} + \epsilon_{ss} - \epsilon_{cs})Ni \equiv -\Delta\epsilon_{in}Ni, \qquad (2.7)$$

where ϵ_{ci}, ϵ_{si}, ϵ_{ss} and ϵ_{cs} are, respectively, the interaction energies of chain segment–tube inside, solvent molecule–tube inside, solvent molecule–solvent molecule, and chain segment–solvent molecule. $\Delta\epsilon_{in}$ represents the inclusion–dissociation energy per chain segment. On the other hand, the dissociated chains interact with each other and with the outside of the tubes. Then, the interaction energy differences $\Delta\epsilon_{cc}$ between the chains, and the interaction energy $\Delta\epsilon_{co}$ between the chain and tube outside are defined by eqn (2.8) and (2.9):

$$\Delta\epsilon_{cc} = \frac{z}{2}(\epsilon_{cc} + \epsilon_{ss} - 2\epsilon_{cs}) \qquad (2.8)$$

$$\Delta\epsilon_{co} = z(\epsilon_{co} + \epsilon_{ss} - \epsilon_{cs} - \epsilon_{so}), \qquad (2.9)$$

where ϵ_{cc}, ϵ_{so}, and ϵ_{co} are, respectively, the interaction energies of chain segment–chain segment, solvent molecule–tube outside, and chain segment–tube outside.

Next, the transition temperature T_c is defined by

$$T_c = \frac{\Delta\epsilon_{in}}{k_B\ln(z-1)} \equiv \frac{\Delta\epsilon_{in}}{\Delta S}, \qquad (2.10)$$

where k_B is the Boltzmann constant and ΔS represents the inclusion–dissociation entropy per chain segment based on the free rotation around covalent bonds. It is to be noted that, when the temperature $T = T_c$, the inclusion energy and the polymer entropy are balanced. According to Flory–Huggins theory, the reduced (*i.e.*, dimensionless) free energy $\bar{F}(p)$ per tube as a function of the inclusion fraction $p = i/N_t$ is:

$$\bar{F}(p) \equiv \frac{F(p)}{N_t k_B T} = \left[\frac{N \Delta S}{k_B} \left(1 - \frac{1}{\bar{T}} \right) - N \Phi_t \left(1 - \frac{\bar{\Theta}_{co}}{\bar{T}} \right) + \ln \frac{Nz}{\Phi_t} \right] p$$

$$+ \frac{N \Phi_t (\alpha - p)^2}{2} \left(1 - \frac{\bar{\Theta}_{cc}}{\bar{T}} \right) - p \ln B(p) + (\alpha - p) \ln(\alpha - p) \qquad (2.11)$$

$$+ (1 - p) \ln(1 - p) + p \ln p,$$

where the terms independent of p have been omitted and $\bar{T} \equiv T/T_c$ is the reduced temperature; $\bar{\Theta}_{co}$ is the reduced temperature defined by $\bar{\Theta}_{co} = \Delta \epsilon_{co} / k_B T_c$, $\bar{\Theta}_{cc}$ is the reduced temperature defined by $\bar{\Theta}_{cc} = \Delta \epsilon_{cc} / k_B T_c$, and $\Phi_t \equiv NN_t / \Omega$. In eqn (2.11), $B(p)$ is defined by:

$$B(p) = \frac{1 - \exp[NC(p)]}{1 - \exp[C(p)]}, \qquad (2.12)$$

where $C(p)$ is given by:

$$C(p) = \left(1 - \frac{1}{\bar{T}} \right) \frac{\Delta S}{k_B} - \Phi_t (\alpha - p) \left(1 - \frac{\bar{\Theta}_{cc}}{\bar{T}} \right). \qquad (2.13)$$

Figure 2.7 shows the $\bar{T}$ dependence of the equilibrium value of p, which has been obtained from the condition $\partial \bar{F}(p)/\partial p = 0$ for different values of N in a good solvent. As N increases, the inclusion–dissociation behavior becomes sharp and approaches a transitional one without hysteresis at the transition temperature $\bar{T} = 1$ or $T = T_c$ in the limit of $N \rightarrow \infty$. On the other hand, the p–$\bar{T}$ curve shown in Figure 2.8 indicates that the inclusion-dissociation transition occurs with a hysteresis loop at $\bar{T} \ll \bar{\Theta}_{cc}$, in the poor solvent region. Figure 2.9 shows the p–α curves for different values of $\bar{\Theta}_{cc}$. The inclusion fraction p increases linearly and saturates with increasing α in the good solvent region. On the other hand, the dissociation transition can occur after the continuous inclusion in the poor solvent.

In summary, the inclusion–dissociation behavior in the complex formation between the molecular nanotubes and the linear polymer chains in solutions was investigated theoretically using the Flory–Huggins lattice model. In a good solvent, the polymer chains are continuously dissociated from or included into the tubes as T or α vary. On the other hand, the inclusion–dissociation transition occurs in a poor solvent with a hysteresis loop. These theoretical results are useful for the control of supramolecular structures formed by nanotubes and polymer chains.

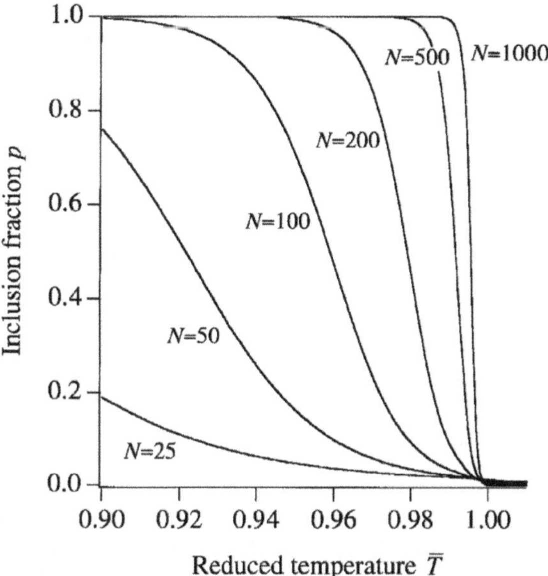

Figure 2.7   The reduced temperature $\bar{T}$ dependence of the equilibrium value of p, which has been obtained from eqn (2.11).[6]
Reproduced by permission of The American Physical Society.

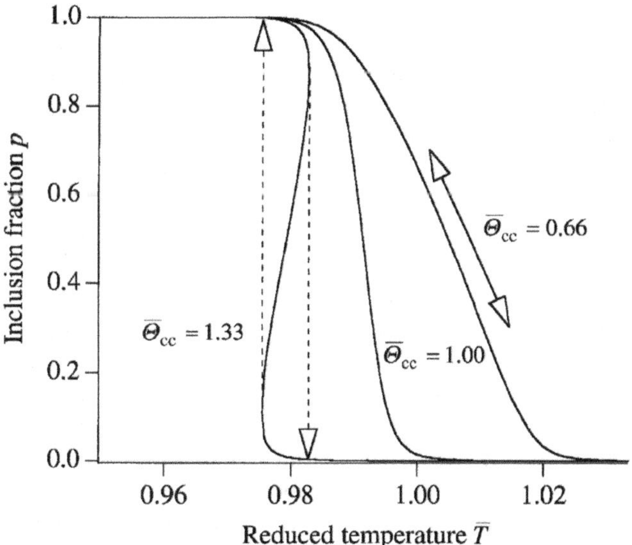

Figure 2.8   The reduced temperature $\bar{T}$ dependence of the inclusion fraction p for different values of the reduced theta temperature $\bar{\Theta}_{cc}$.[6]
Reproduced by permission of The American Physical Society.

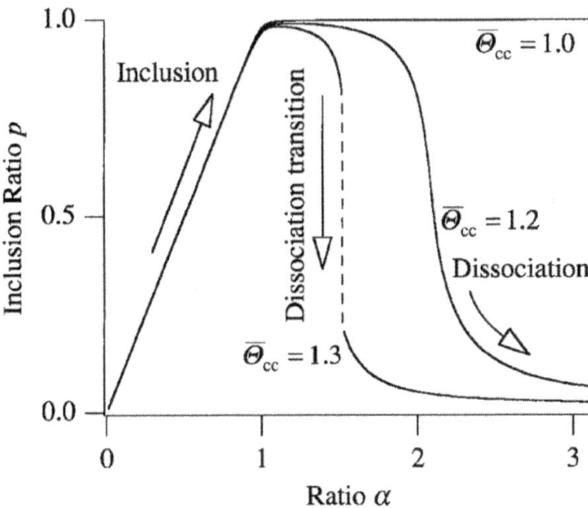

Figure 2.9 The dependence of the inclusion fraction p on the ratio $\alpha = N_c/N_t$ for different values of the reduced theta temperature $\bar{\Theta}_{cc}$.[6]
Reproduced by permission of The American Physical Society.

2.3 Sliding Transition of Rings and Tubes on a Backbone Polymer

If an axis polymer string is a block copolymer consisting of some components with different interactions with rings or tubes, the polyrotaxane may indicate sliding transition of the rings or tubes on the backbone string polymer. Such a polyrotaxane with a backbone block copolymer was actually synthesized by Yui *et al.*[7] For example, rings are accumulated on a component of the block copolymer with stronger attractive interaction at low temperature; however, they are distributed randomly over the block copolymer at high temperature.[8] The sliding transition of the rings or tubes on polyrotaxane may be applicable to a molecular switch or an actuator of nano-machines.

An inclusion complex of the molecular nanotube and a block copolymer comprising linear polymer chains of two types, A and B, is shown in Figure 2.10.[9] Both ends of the block copolymer are terminated with bulky ends of large substitutional groups so that the block copolymer cannot dissociate from the nanotube. If the A chain has inclusion energy and entropy larger than the B chain, the nanotube includes the A chain at low temperature or the B chain at high temperature. In other words, the nanotube switches back and forth between the A and B chains as the temperature varies. Thus, we call such a polyrotaxane having the block copolymer backbone a "switching complex". In addition to the conformational entropy and the tube–chain interaction energy, the property of miscibility between A and B chains is another important factor dominating the switching behavior. This effect is incorporated into the following theory by using the Flory–Huggins lattice model.

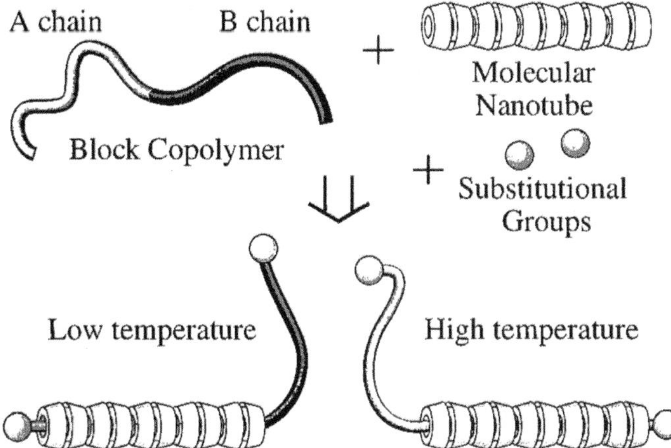

Figure 2.10 A switching complex formation of a molecular nanotube and a block polymer.[9]
Reproduced by permission of The American Physical Society.

A lattice model similar to those in the previous sections is assumed, *i.e.*, that a molecular nanotube and an A–B block copolymer are composed of segments with the same size as a solvent molecule occupying one site on the lattice.[9] In this system, let N_c be the number of switching complexes. The inside of the tube is filled with either the A or B chain segment, so that the tube segment also occupies one lattice site together with the chain segment. For simplicity, it is assumed that the tube is fully rigid and has a longitudinal length equal to the contour length of the A or B chain, namely, each of the tubes, and the A and B chains consists of N segments occupying N connected lattice points. Therefore, the nanotube can move by N segments on the block copolymer.

When n is the number of B segments included in the nanotube and corresponds to the displacement of the tube from the full inclusion state with the A chain, the free-energy difference of the switching complex is evaluated from the full inclusion state with the A chain as follows:

$$F_s(n, T) = n[(z - 2)(\epsilon_{ai} - \epsilon_{as} - \epsilon_{bi} + \epsilon_{bs}) - T(S_a - S_b)] \equiv n[\Delta\epsilon_s - T\Delta S_s], \quad (2.14)$$

where z is the coordination number of the lattice, ϵ_{ai}, ϵ_{as}, ϵ_{bi}, and ϵ_{bs} are, respectively, the interaction energy of A segment–tube inside pairs, A segment–solvent molecule pairs, B segment–tube inside pairs, and B segment–solvent molecule pairs. In eqn (2.14), S_a and S_b represent the conformational entropy per segment of the A and B chains, respectively, based on the free rotation around covalent bonds. Therefore, $\Delta\epsilon_s$ and ΔS_s indicate the differences in inclusion energy and entropy between A and B segments. It is to be noted that when $T = T_s \equiv \Delta\epsilon_s/\Delta S_s$, inclusion forces toward the A and B chains are balanced.

On the other hand, linear polymer chains extruding from the tubes interact with each other. With a mean-field approximation, the interaction energy $E_i(n,T)$ is given by eqn (2.15):

$$E_i(n, T) = -\Phi[np\Delta\epsilon_{aa} + n(1-p)\Delta\epsilon_{ab} + (N-n)(1-p)\Delta\epsilon_{bb} + (N-n)p\Delta\epsilon_{ab}], \tag{2.15}$$

where p is the switching ratio defined by $p = \bar{n}/N$ with the average number $\bar{n}$ of n, and $\Phi = NN_c/\Omega$ denotes the volume fraction of polymer chains of each type or tubes. In eqn (2.15), $\Delta\epsilon_{aa}$, $\Delta\epsilon_{ab}$, and $\Delta\epsilon_{ab}$ are defined by:

$$\Delta\epsilon_{aa} = (z-2)(\epsilon_{aa} + \epsilon_{ss} - 2\epsilon_{as}) \tag{2.16}$$

$$\Delta\epsilon_{bb} = (z-2)(\epsilon_{bb} + \epsilon_{ss} - 2\epsilon_{bs}) \tag{2.17}$$

$$\Delta\epsilon_{ab} = (z-2)(\epsilon_{ab} + \epsilon_{ss} - \epsilon_{as} - \epsilon_{bs}), \tag{2.18}$$

where ϵ_{aa}, ϵ_{bb}, ϵ_{ab} and ϵ_{ss} are, respectively, the interaction energy of A segment–A segment, B segment–B segment, A segment–B segment, and solvent molecule–solvent molecule. The total free energy $F(n,T)$ is given by $F(n,T) = F_s(n,T) + E_i(n,T)$. Accordingly, when the tube moves by a segment towards the B chain, a change in the free energy is given by:

$$\Delta F = \frac{dF(n, T)}{dn} = k_B T_s[\Delta\bar{\epsilon}_s(1 - \bar{T}) + \Phi[(2p - 1)\bar{\Theta}_m - \bar{\Theta}_s]]. \tag{2.19}$$

Here we introduce a reduced energy $\Delta\bar{\epsilon}_s = \Delta\epsilon_s/(k_B T_s)$ and reduced temperatures, $\bar{T} = T/T_s$, $\bar{\Theta}_m = (2\Delta\epsilon_{ab} - \Delta\epsilon_{aa} - \Delta\epsilon_{bb})/(2k_B T_s)$ and $\bar{\Theta}_s = (\Delta\epsilon_{aa} - \Delta\epsilon_{bb})/(2k_B T_s)$. $\bar{\Theta}_m$ and $\bar{\Theta}_s$ represent, respectively, the miscibility and the difference in solubility between A and B chains.

The partition function P of a switching complex is given by:

$$P = \sum_{n=0}^{N} K^n = \frac{1 - K^{N+1}}{1 - K}, \tag{2.20}$$

where

$$K = \exp\left[-\frac{\Delta F}{k_B T}\right]. \tag{2.21}$$

Therefore, we obtain the switching ratio p of the complex as:

$$p(K) = \frac{1}{NP}\sum_{n=0}^{N} nK^n = \frac{-K}{1 - K^{N+1}}\left(K^N - \frac{1 - K^N}{N(1 - K)}\right). \tag{2.22}$$

Since K depends on p, as shown in eqn (2.19) and (2.21), we determine p from eqn (2.22) self-consistently.

Figure 2.11 shows the $\bar{T}$ dependence of p for different values of N at $\bar{\Theta}_m = 0$, *i.e.*, in the case of miscible block copolymers. As N increases, the switching behavior becomes sharp and approaches a transitional behavior without hysteresis at the transitional temperature $\bar{T} = 1$ in the limit of $N \to \infty$.

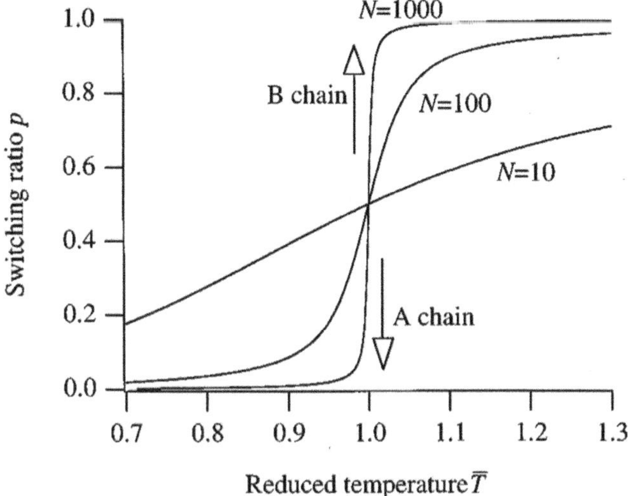

Figure 2.11   The reduced temperature $\bar{T}$ dependence of p for different values of N in the miscible block copolymers.[9]
Reproduced by permission of The American Physical Society.

Figure 2.12a and b show the $\bar{T}$ dependence of p for miscible block copolymers ($\bar{\Theta}_m \geq 0$) and immiscible block copolymers ($\bar{\Theta}_m < 0$), respectively. Figure 2.12a shows that as $\bar{\Theta}_m$ increases, the nanotubes shift gradually with varying temperature on miscible block copolymers. Conversely, the p–$\bar{T}$ curve shown in Figure 2.12b indicates that the switching transition occurs with a hysteresis in immiscible block copolymers. Figure 2.13 shows the p–$\bar{T}$ curve for different values of $\bar{\Theta}_s$ at $\bar{\Theta}_m < 0$. It is seen that the transition temperature strongly depends on $\bar{\Theta}_s$, which is the solubility difference between A and B chains.

If the nanotube includes a recurrent block copolymer comprising linear polymer chains of four different types (A, B, C, D) with the same contour length L, as shown in Figure 2.14a, the switching behavior yields a fascinating feature. Each polymer chain has the inclusion energy and entropy of $(-\epsilon_a, S_a)$, $(-\epsilon_b, S_b)$, $(-\epsilon_c, S_c)$, and $(-\epsilon_d, S_d)$, respectively. Here, it is assumed that $\epsilon_b > \epsilon_a \gg \epsilon_d > \epsilon_c$ and $S_a \approx S_b \gg S_c \approx S_d$, namely, that the A and B chains are more flexible and energetically more stable than the C and D chains, and the B and D chains have larger inclusion energies than the A and C chains, respectively. When the nanotube with length L includes the block copolymer, the free-energy map along the contour of the block copolymer is given as Figure 2.14b and changes drastically with varying temperature. At a low temperature, including the A chain at the end of the block copolymer, the tube passes the A chain and stays at the B chain. As temperature rises, the free-energy level at the B chain becomes equal to that at the C chain at the temperature $T_{bc} = (\epsilon_b - \epsilon_c)/(S_b - S_c)$, so that the tube moves toward the C chain, passes the C chain, and stays at the D chain. It is to be noted that the tube does not travel backward, since the B chain is more stable than the A

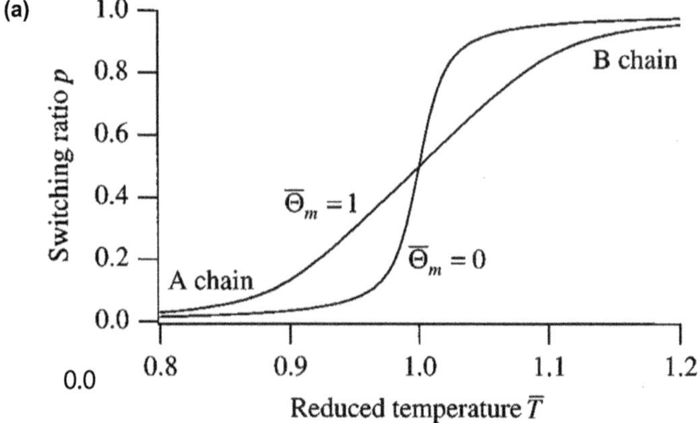

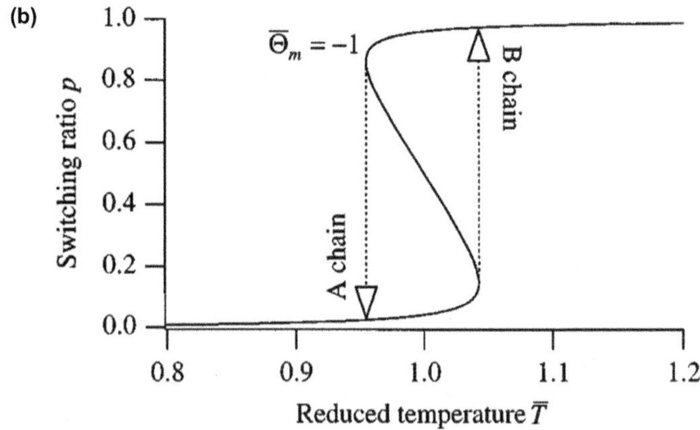

Figure 2.12 The reduced temperature $\overline{T}$ dependence of p for miscible block copolymers (a) ($\overline{\Theta}_m \geq 0$), and immiscible block copolymers (b) ($\overline{\Theta}_m < 0$).[9] Reproduced by permission of The American Physical Society.

chain at any temperature. When temperature falls to $T_{da} = (\epsilon_d - \epsilon_a)/(S_d - S_a)$, the free-energy level at the D chain becomes equal to that at the A chain, and the tube at the D chain moves to the B chain. That is, the temperature undulation between T_{bc} and T_{da} moves the tube stepwise in one direction on the block copolymer, as schematically shown in Figure 2.14c. In other words, we can control the displacement of the tube on the block polymer, which can be regarded as a molecular motor "nano-rail". The amplitude of the recurrent free energy, which produces the driving force of the nano-rail, is in proportion to the length L of the nanotube. Therefore, if L is long enough, the nanotube does not diffuse in either direction by thermal agitation on the block copolymer, unlike small cyclic molecules such as cyclodextrins. The transitional behavior is a common feature of supramolecular systems comprising nanotubes and linear polymer chains.

To summarize, the switching behavior of the molecular nanotube in the switching complexes in solutions was investigated theoretically using the

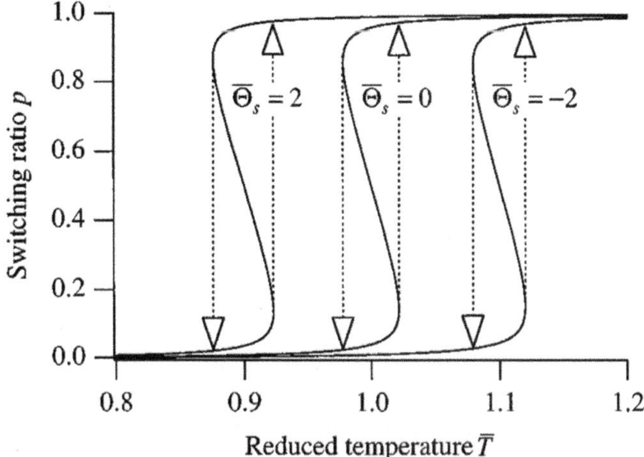

Figure 2.13 The reduced temperature $\overline{T}$ dependence of the switching ratio p for different values of $\overline{\Theta}_s$ at $\overline{\Theta}_m < 0$.[9]
Reproduced by permission of The American Physical Society.

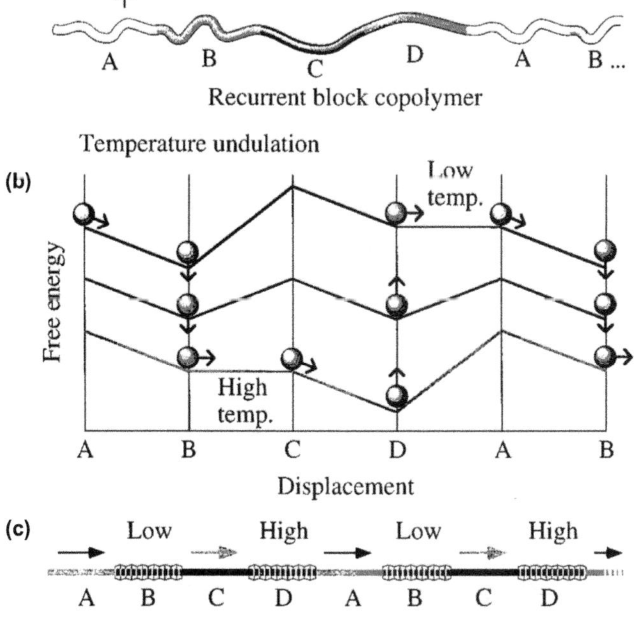

Figure 2.14 (a) Nanotube and recurrent block copolymer comprising linear polymer chains of four different types. (b) Schematic free-energy map along the block copolymer for different temperatures. The ball represents the location of the nanotube on the block copolymer. (c) Stepwise motion of the nanotube, induced by temperature undulation, on the block copolymer in one direction.[9]
Reproduced by permission of The American Physical Society.

Flory–Huggins lattice model. For miscible block copolymers, the nanotubes continuously move back and forth between A and B chains as the temperature varies. On the other hand, the switching transition occurs for immiscible block copolymers with a hysteresis loop. When a recurrent block copolymer made of four kinds of polymer chains is used, one can realize the molecular motor where the displacement of the nanotube is controllable by the temperature undulation. These theoretical results suggest high function of the supramolecular system consisting of the molecular nanotubes and the block copolymers.

As mentioned in Chapter 1, mobile rings in a polyrotaxane have distribution or arrangement entropy. Therefore, the entropic force arises from the translational entropy of mobile rings when an external force yields the heterogeneous distribution of rings, such as accumulation. Sevick and Williams theoretically investigated the thermomechanical responses of various polyrotaxane or rotaxane systems, where rings in the polyrotaxane behave as a molecular piston or shock absorber.[10] Figure 2.15 shows such a

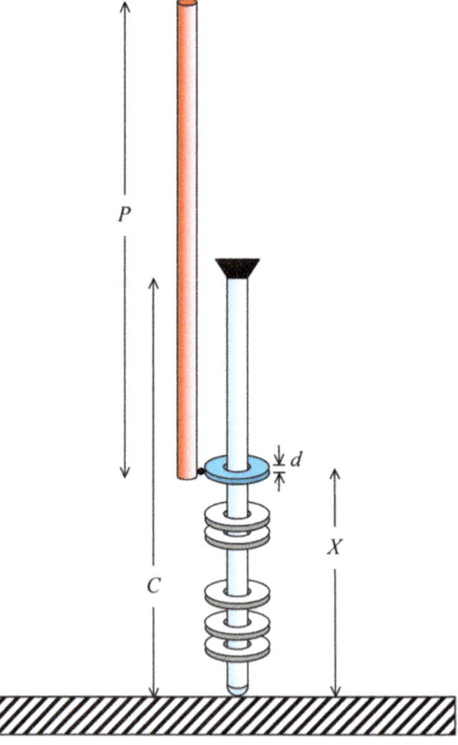

Figure 2.15   Schematic of a molecular shock absorber constructed from a piston attached to a ring of a rotaxane molecule that is surface-grafted, where P and C are the lengths of the piston and rotaxane, respectively, d is the thickness of rings, and X the distance of the ring attached to the piston from the surface.[10]
Reprinted with permission from Langmuir. Copyright (2010) American Chemical Society.

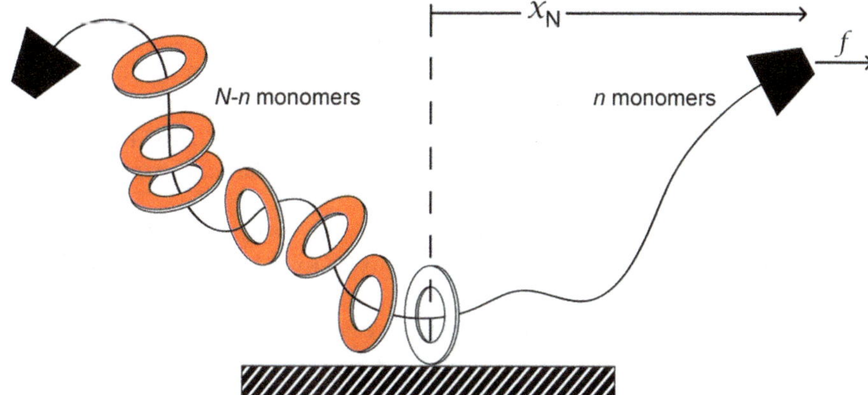

Figure 2.16 A schematic of a polyrotaxane, stoppered at both ends to retain the rings. The control-ring (white) is fixed to the surface at the origin, and at any instant in time the control-ring separates the chain into two components: a ring-less sub-chain of n monomers and a ring-threaded sub chain of N-n monomers. A force, f, is applied to the end of the ring-less sub-chains. Monomers can pass through the control-ring, allowing for an exchange of monomers of the sub-chains; but as the rings are volume excluding, they are unable to trespass the control-ring and must remain on the ring-threaded sub-chain, where x_N is the distance between the control ring and chain end.[11]
Reprinted with permission from Macromolecules. Copyright (2013) American Chemical Society.

system consisting of a rod-like polyrotaxane connected to a piston and to a surface. The free energy was calculated exactly from the translational entropy of the mobile rings, which yielded the resistance to the external force on the piston, as expected. In some cases, the polyrotaxane exhibited a sudden transition to a tilted state when rings are compressed. Subsequently, they treated the effect of mobile rings on the conformation of a single polyrotaxane molecule with a control-ring fixed on the surface, as shown in Figure 2.16.[11] It was found that a small force applied resulted in dramatic stretching of the backbone string. In particular, if the mobile rings are located asymmetrically around the fixed ring, a small force can extend the backbone string fully, suggesting a yield force below which chain extension is minimal. In contrast, the modulus at large forces arises from the conformational entropy of the backbone string, where the entropy of rings becomes negligible. The predictions are in qualitative agreement with the experimental results.[12]

References

1. A. Harada and M. Kamachi, *Macromolecules*, 1990, **23**, 2821.
2. A. Harada, J. Li and M. Kamachi, *Nature*, 1993, **364**, 516.
3. M. Doi and S. F. Edwards, *The Theory of Polymer Dynamics*, Oxford Press, 1990.

4. Y. Okumura, K. Ito and R. Hayakawa, *Polym. Adv. Technol.*, 2000, **11**, 815.
5. A. Harada, J. Li and M. Kamachi, *Macromolecules*, 1994, **27**, 4538.
6. Y. Okumura, K. Ito and R. Hayakawa, *Phys. Rev. Lett.*, 1998, **80**, 5003.
7. H. Fujita, T. Ooya and N. Yui, *Macromol. Chem. Phys.*, 1999, **200**, 706.
8. H. Fujita, T. Ooya and N. Yui, *Macromolecules*, 1999, **32**, 2534.
9. Y. Okumura, K. Ito and R. Hayakawa, *Phys. Rev. E, Rapid. Commun.*, 1999, **59**, R3823.
10. E. M. Sevick and D. R. M. Williams, *Langmuir*, 2010, **26**, 5864.
11. M. B. Pinson, E. M. Sevick and D. R. M. Williams, *Macromolecules*, 2013, **46**, 4191.
12. K. Kato, T. Yasuda and K. Ito, *Macromolecules*, 2013, **46**, 310.

Scattering Studies of Polyrotaxane and Slide-ring Materials

3.1 Structure Analysis of Polyrotaxane in Solution

Polyrotaxane is a topologically interlocked supramolecular assembly where cyclic molecules are threaded on a linear polymer. The necklace-like molecular architecture has been utilized to create various kinds of functional materials such as insulated molecular wires consisting of cyclodextrins and conjugated polymers,[1] molecular nanotubes formed by cross-linking cyclic molecules on polyrotaxanes,[2] and drug–polyrotaxane conjugates as supramolecular structured drug carriers.[3] The molecular structure of polyrotaxane has been investigated by small-angle neutron scattering (SANS) techniques.[4] SANS has been a powerful tool to measure the structure of polymeric systems[5] since it was used to observe the random oil conformation of polymers in the melt.[6]

Mayumi *et al.* have studied the conformation of polyrotaxane composed of poly(ethylene glycol) (PEG), and α-cyclodextrin (CD) by SANS.[7] Figure 3.1 shows the scattering function $I(Q)$ for hydroxylpropylated polyrotaxane (HyPR) and PEG in dilute solutions. Q is the magnitude of the scattering vector:

$$Q = \frac{4\pi}{\lambda} \sin\left(\frac{\theta}{2}\right), \tag{3.1}$$

where λ is the neutron wavelength and θ is the scattering angle. $I(Q)$ for the PEG solution shows a power–law behavior $I(Q) \sim Q^{-5/3}$ in the high-Q regime, which is consistent with the fractal dimension of flexible polymer chains in a

Monographs in Supramolecular Chemistry No. 15
Polyrotaxane and Slide-Ring Materials
By Kohzo Ito, Kazuaki Kato and Koichi Mayumi
© Kohzo Ito, Kazuaki Kato and Koichi Mayumi 2016
Published by the Royal Society of Chemistry, www.rsc.org

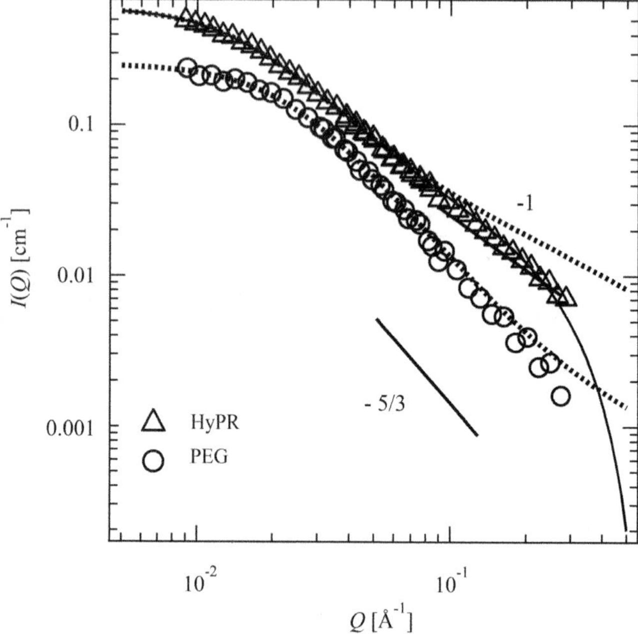

Figure 3.1   Scattering intensity *vs* scattering vector Q for HyPR and PEG in dilute solutions. The dashed and solid curves are derived from the form factor for the worm-like chain model [eqn (3.2)] and that for the modified worm-like chain model, taking account of the polymer radius [eqn (3.7)], respectively.[4]
Reprinted from Polymer, 51, 959–967, Copyright (2010), with permission from Elsevier.

good solvent.[8] For HyPR, the slope of $I(Q)$ in the high-Q region is between $-5/3$ and -1. This suggests a more rigid structure compared with PEG. The dotted lines in Figure 3.1 are the fitting results from the worm-like chain model. The form factor for the worm-like chain model is given by:[9]

$$P_{KP}(Q) = [(1 - \chi(Q,L))P_{Debye}(Q,L) + \chi(Q,L)P_{Rod}(Q,L)]\Gamma(Q,L), \qquad (3.2)$$

where L is the contour length of the polymer chain, and $\Gamma(Q,L)$ is the additive factor calculated by Yoshizaki and Yamakawa.[9] $P_{Debye}(Q,L)$ is the scattering function for random coils:

$$P_{Debye}(Q) = 2(Q^2 R_g^2)^{-2}(\exp(-Q^2 R_g^2) + Q^2 R_g^2 - 1) \qquad (3.3)$$

The radius of gyration for the polymer R_g is a function of the persistence length l_p:

$$R_g^2 = \frac{L l_p}{3} - l_p^2 + \frac{2 l_p^3}{L} - \frac{2 l_p^4}{L^2}\left(1 - \exp\left(-\frac{L}{l_p}\right)\right). \qquad (3.4)$$

$P_{\text{Rod}}(Q,L)$ in eqn (3.2) is the form factor for a rod with the length L:

$$P_{\text{Rod}}(Q) = \frac{2}{QL} Si(QL) - \sin^2\left(\frac{QL}{2}\right), \qquad (3.5)$$

where Si is the sine integral. The factor $\chi(Q,L)$ in eqn (3.2) is an increasing function with Q, which represents the balance between $P_{\text{Debye}}(Q,L)$ and $P_{\text{Rod}}(Q,L)$.

$$\chi(Q,L) = \exp(-(\pi R_g^2 Q/2L)^{-5}). \qquad (3.6)$$

As shown in Figure 3.1, $I(Q)$ for the PEG solution is fitted well with the worm-like chain model $P_{\text{KP}}(Q,L)$, and the obtained l_p is 11 Å. In the case of the HyPR solution, $I(Q)$ deviates from the dotted line from the worm-like chain model in the high-Q regime. This is because HyPR has a thicker diameter than usual polymers due to the necklace-like structure. For the structural analysis of HyPR, $P_{\text{Rod}}(Q,L)$ in eqn (3.2) should be replaced by the form factor of a cylinder $P_{\text{Cyl}}(Q, L, R)$ with a diameter R:

$$P_{\text{PR}}(Q) = [(1 - \chi(Q, L))P_{\text{Debye}}(Q, L) + \chi(Q, L)P_{\text{Cyl}}(Q, L)]\Gamma(Q, L) \qquad (3.7)$$

$$P_{\text{Cyl}}(Q,L,R) = \int_0^{\pi/2} \left\{ \frac{2J_1(QR \cdot \sin\alpha)}{QR \cdot \sin\alpha} \frac{\sin\left(\frac{QL}{2}\cos\alpha\right)}{\frac{QL}{2}\cos\alpha} \right\}^2 \sin\alpha \, d\alpha. \qquad (3.8)$$

Here, $J_1(x)$ is the first-order cylindrical Bessel function. The scattering function $I(Q)$ for the HyPR solution is described well by the modified worm-like chain model with the radius $R = 7$ Å, the size of CD (the solid line in Figure 3.1). By the model fitting, the persistence length of HyPR is determined to be 44 Å, which is four times as large as that of PEG without CDs. This means that covering PEG with CDs stretches the axial polymer chain. From eqn (3.4) and the persistence length l_p of HyPR, the radius of gyration R_g is estimated as 200 Å, which is close to R_g measured by static light scattering.[10] R_g of polyrotaxane changes with the filling ratio of CD on PEG. Fleury *et al.* have synthesized polyrotaxanes having different amounts of CD per chain from 3 to 125, and have measured R_g values of the polyrotaxanes by means of SANS.[11] They have found that R_g increases drastically with the number of CD on PEG. Yamada *et al.* also have performed small-angle X-ray scattering (SAXS) experiments on dilute polyrotaxane solutions.[12] The flexibility of polyrotaxane with various numbers of CDs is estimated quantitatively by the worm-like cylinder model. From the SAXS analysis and viscosity measurements on the polyrotaxane solutions, they have concluded that the chain stiffness increases with the number of CDs, and that the contour length of polyrotaxane becomes shorter than the template PEG chain.

The more detailed structure of polyrotaxane in solutions, especially the distribution of CDs on PEG, has been investigated by contrast variation SANS techniques.[13] One of the advantages of neutron scattering is that the hydrogen/deuterium replacement allows the isotope labeling, since the coherent scattering length for the neutron is quite different for hydrogen and deuterium.[5] The contrast variation enables us to elucidate the detailed structure of each part in multicomponent complex systems, including polymer/nanoparticle complexes,[14] polymer aggregates with block copolymers,[15] inorganic crystals with polymers,[16] and so on. Mayumi *et al.* have conducted contrast variation SANS experiments on polyrotaxane solutions using hydrogenated polyrotaxane (h-PR) with hydrogenated PEG and deuterated PR (d-PR) with deuterated PEG.[13] The scattering function for PR solutions is expressed as a sum of three partial scattering functions $S_{ij}(Q)$ (Figure 3.2):

$$I(Q) = \Delta\rho_{\mathrm{C}}{}^{2}S_{\mathrm{CC}}(Q) + 2\Delta\rho_{\mathrm{C}}\Delta\rho_{\mathrm{P}}S_{\mathrm{CP}}(Q) + \Delta\rho_{\mathrm{P}}{}^{2}S_{\mathrm{PP}}(Q), \tag{3.9}$$

where $\Delta\rho_{i}$ is the difference of the scattering length density between a component i (C for CD or P for PEG) and a solvent. The self-terms, $S_{\mathrm{CC}}(Q)$ and $S_{\mathrm{PP}}(Q)$, correspond to the structure of CD and the conformation of PEG, respectively, and the cross-term $S_{\mathrm{CP}}(Q)$ denotes the pair correlation of CD and PEG. To determine the partial scattering functions, $I(Q)$ values for PR solutions with different scattering contrasts were measured, as shown in

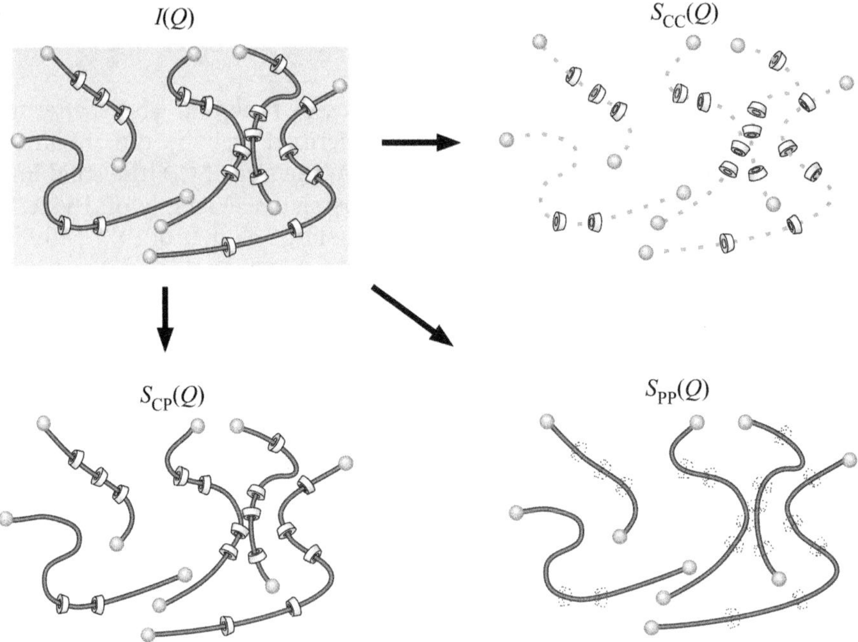

Figure 3.2   Schematic representations showing the decomposition of scattering intensity $I(Q)$ for polyrotaxane solutions into three partial scattering functions: self-term of PEG, $S_{\mathrm{PP}}(Q)$; self-term of CD, $S_{\mathrm{CC}}(Q)$; cross-term of PEG and CD, $S_{\mathrm{CP}}(Q)$.[13]
Reprinted with permission from Macromolecules. Copyright (2009) American Chemical Society.

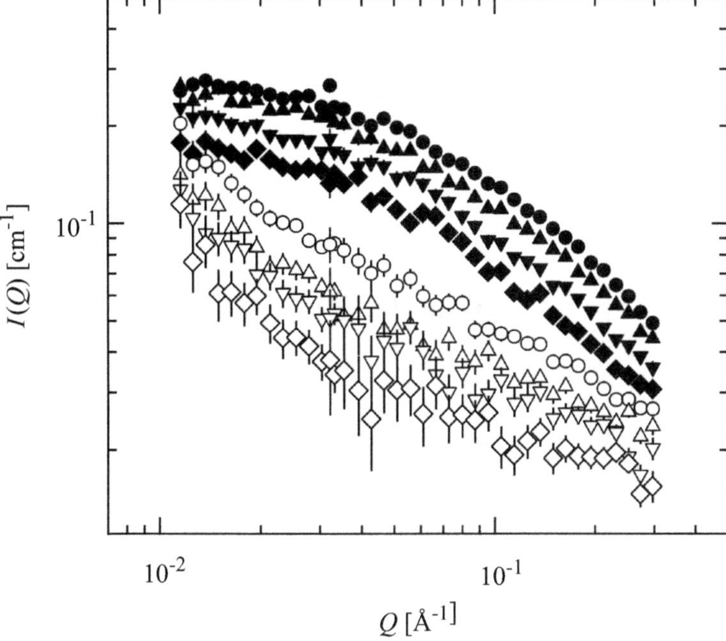

Figure 3.3 Scattering functions of h-PR (solid symbols) and d-PR (open symbols) in mixtures of dimethylsulfoxide-d_6 (DMSO-d_6) and DMSO. Circles ($\bullet$, $\circ$), upward triangles ($\blacktriangle$, $\triangle$), downward triangles ($\blacktriangledown$, $\triangledown$), and diamonds ($\blacklozenge$, $\lozenge$) correspond to $\phi_{\mathrm{DMSO}\text{-}d6} = 1.0$, 0.95, 0.90 and 0.85, respectively, where $\phi_{\mathrm{DMSO}\text{-}d6}$ is the volume fraction of DMSO-d_6 in the solvents.[13] Reprinted with permission from Macromolecules. Copyright (2009) American Chemical Society.

Figure 3.3. They used h-PR and d-PR in DMSO-d_6/DMSO mixtures with four different fractions of DMSO-d_6. By the singular decomposition analysis based on eqn (3.9), the scattering functions for the h-PR and d-PR solutions are decomposed into the partial scattering functions $S_{ij}(Q)$ (Figure 3.4). The positive cross-term $S_{\mathrm{CP}}(Q)$ indicates the mechanically interlocked connection between CD and PEG. The self-term of CD, $S_{\mathrm{CC}}(Q)$, shows almost the same Q dependence as the self-term of PEG, $S_{\mathrm{PP}}(Q)$. This means that the CDs in the polyrotaxane are distributed randomly along the entire PEG chain. Endo *et al.* have calculated the form factor for the cyclic molecules threaded onto the polymer chains as a function of the probability ϕ_{cc} that two cyclic molecules are successively placed on the chain.[17] By fitting the partial scattering function S_{CC} with the form factor, ϕ_{cc} is evaluated to be 0.25, which is close to the average filling ratio of CD in the polyrotaxane, *i.e.*, 0.27. This suggests the random arrangement of CDs on PEG.

The dispersion of CDs in polyrotaxane solutions depends on the concentration of polyrotaxane and the chemical modification of the CDs. Travelet *et al.* have conducted SANS measurements on the aggregation structure of polyrotaxane in concentrated solutions.[18] They have reported that a polyrotaxane composed of CDs and PEG results in nano-cylinder aggregation

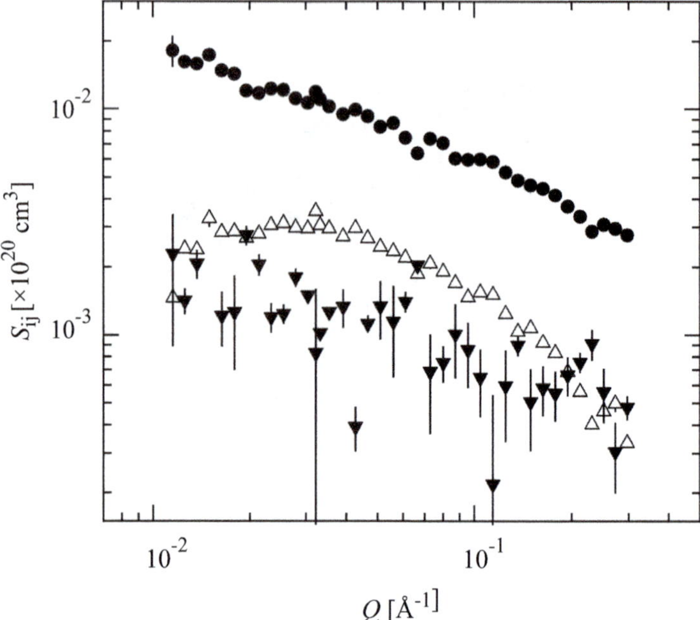

Figure 3.4   Partial scattering functions of CD, S_{CC} (●); PEG, S_{PP} (▼); and CD-PEG, S_{CP} (△).[13]
Reprinted with permission from Macromolecules. Copyright (2009) American Chemical Society.

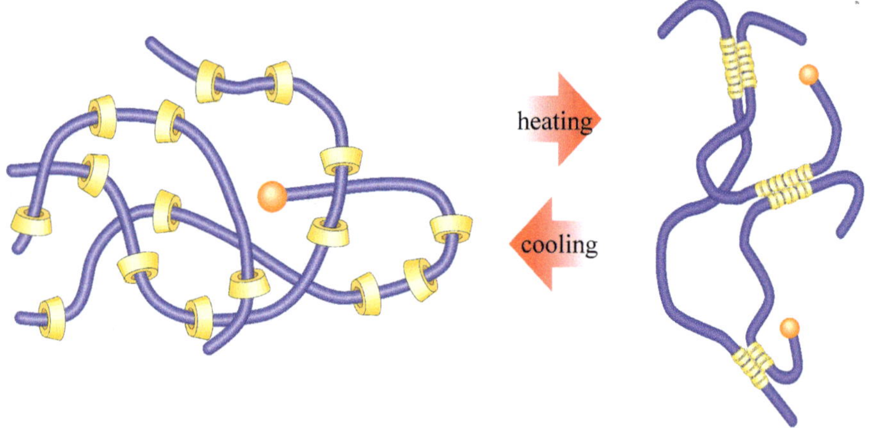

Figure 3.5   Schematic illustration showing the temperature-induced sol–gel transition of methylated polyrotaxane in solution.[19]
Adapted with permission from Macromolecules. Copyright (2006) American Chemical Society.

of CD in concentrated DMSO solutions (about 20 wt%). Karino *et al.* have investigated the sol–gel transition and thermosensitivity of a polyrotaxane consisting of methylated CDs and PEG by means of SANS.[19] The methylated polyrotaxane solution exhibited a sol–gel transition at 60 °C, resulting from the hydrophobic interaction between the methylated CDs (Figure 3.5). At high temperatures, the methylated CDs form crystal-like aggregates,[20] which lead to the gel formation.

3.2 Deformation Mechanism of Slide-ring Gels

Polymer gels, which are cross-linked polymer networks containing solvent, have been expected to be used for biomedical applications, such as artificial skins/vessels and joint prosthesis, because of their high biocompatibility.[21] A significant problem of polymer gels for these applications was their low mechanical strength. In conventional chemical gels cross-linked by covalent bonds, the cross-links are distributed randomly in the polymer network. When chemical gels are deformed, the inhomogeneity is enhanced, as shown in Figure 3.6, and internal stress concentrates on highly cross-linked chains (short strands), which leads to the weak mechanical toughness of the chemical gels. The inhomogeneous network structure of the chemical gels has been investigated by SANS, SAXS and light scattering (LS).[22] Under uniaxial deformation, chemical gels exhibit an anisotropic scattering pattern elongated in the stretching direction, called the "abnormal butterfly pattern". The origin of the abnormal butterfly pattern is the inhomogeneity in deformed gels. Onuki has proposed a model showing that the concentration fluctuation of the frozen heterogeneity gives rise to the abnormal butterfly pattern.[23]

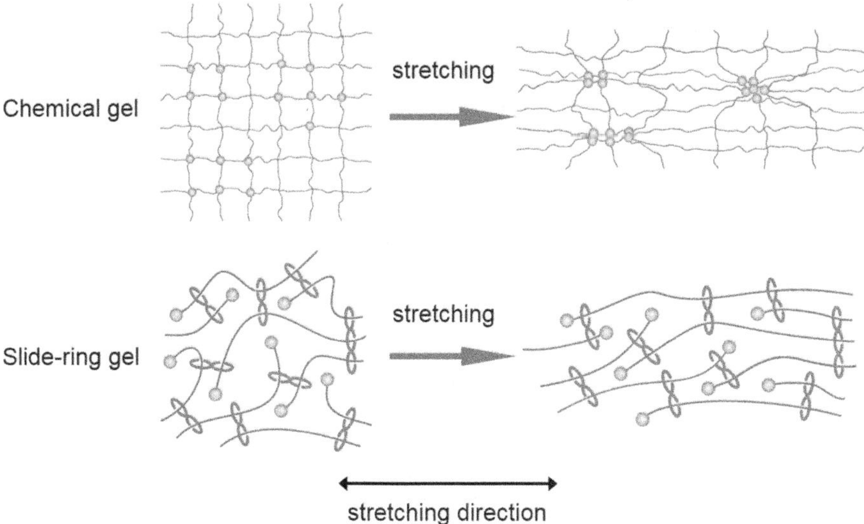

Figure 3.6 Schematic representation showing the deformation mechanism for chemical and slide-ring gels.[26]
Reprinted with permission from Macromolecules. Copyright (2005) American Chemical Society.

Recently, sophisticated chemical designs have been found to improve the mechanical strength of polymer gels.[21,22] Okumura and Ito have developed highly stretchable polymer gels with movable cross-links, called "slide-ring gels".[24,25] By connecting cyclic molecules on different polyrotaxanes, polymer chains are cross-linked *via* slidable cross-links composed of two rings. The figure-of-eight cross-links slide on the polymer chains and behave like pulleys to homogenize the strand length between the cross-links and suppress the stress concentration under deformation. The mobility of the cross-links causes high deformability of the slide-ring gels. The mechanical properties are introduced in the Chapter 4. In this section, we focus on SANS and SAXS analysis of the deformed slide-ring gels to confirm the pulley effect.

Karino *et al.* have carried out SANS measurements on stretched slide-ring gels with different cross-linker concentrations (CX10: 1.0 wt%, CX20: 2.0 wt%).[26] Figure 3.7 represents the two-dimensional (2-D) scattering patterns for the slide-ring gels at various stretching ratios. Interestingly, in the case of the low cross-linker concentration, the elliptical pattern is extended in the direction perpendicular to the stretching. This type of scattering pattern is called the normal butterfly pattern, which is observed in uncross-linked polymer solutions in flow fields.[27] The normal butterfly pattern is derived from the thermal fluctuations in uniformly deformed polymer networks.[23] The appearance of the normal butterfly pattern

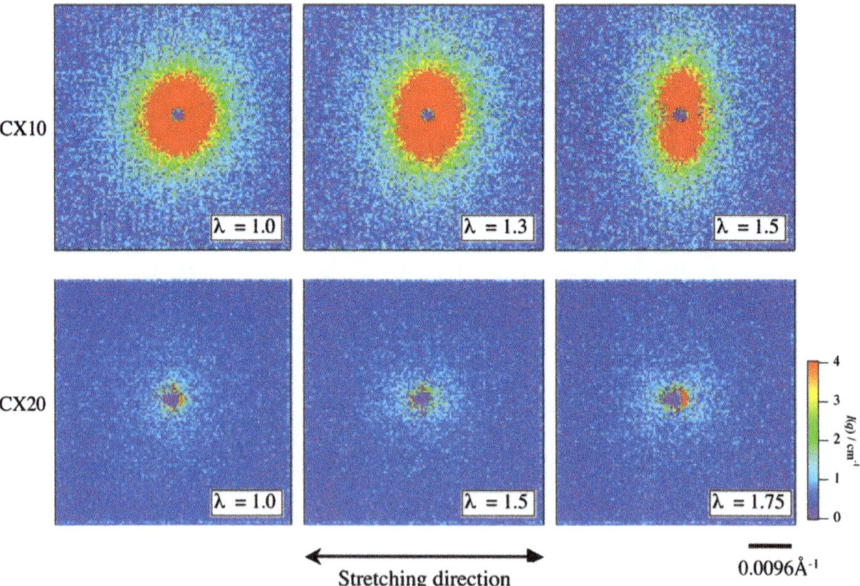

Figure 3.7 Two-dimensional (2-D) SANS patterns of slide-ring gels with different cross-linker concentrations (CX10: 1.0 wt%, CX20: 2.0 wt%).[26]
Reprinted with permission from Macromolecules. Copyright (2005) American Chemical Society.

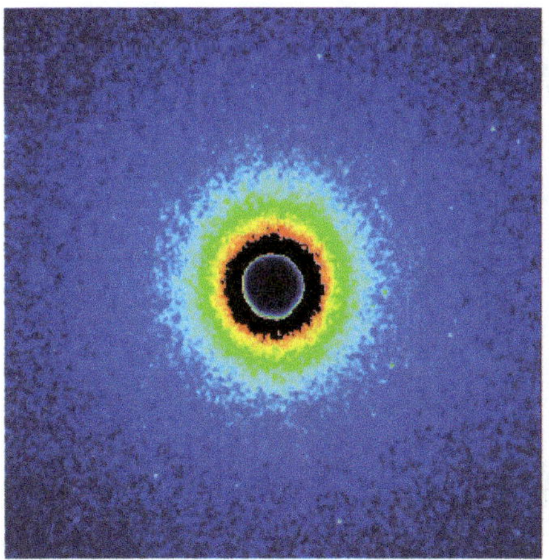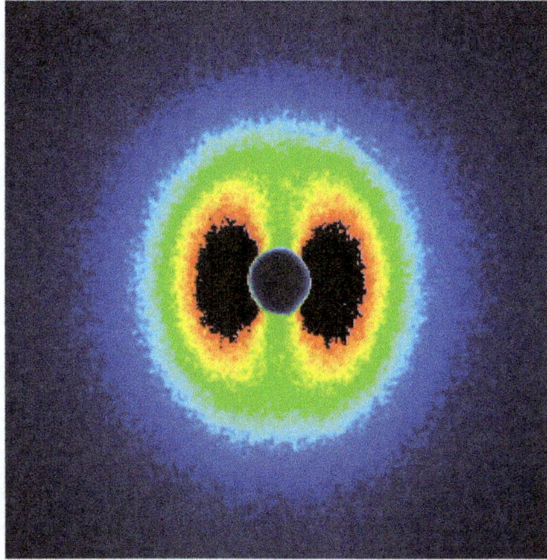

Figure 3.8 SAXS patterns of slide-ring gels with NaOH (left) and NaCl (right) aqueous solutions on uniaxial deformation two times in length in the horizontal axis. NaOH aqueous solution is a good solvent for polyrotaxane, while polyrotaxane forms aggregates in NaCl aqueous solutions.[25]
Reprinted with permission from Polymer Journal. Copyright (2007) Macmillan Publishers Ltd.

suggests that the spatial inhomogeneity is suppressed for the slide-ring gels under deformation, which results from the pulley effect (Figure 3.6). On the other hand, the slide-ring gel with the high cross-linker concentration (CX20) shows an abnormal butterfly pattern, like conventional chemical gels with fixed cross-links. The pulley effect does not work for highly cross-linked slide-ring gels. The efficiency of the pulley effect also depends on solvent quality. Shinohara *et al.* have performed SAXS experiments on deformed slide-ring gels with various kinds of solvent.[28] As shown in Figure 3.8, the slide-ring gels in good solvent exhibits the normal butterfly pattern with stretching. The abnormal butterfly pattern for the slide-ring gels in poor solvent suggests that the aggregation of cyclic molecules suppresses the sliding of the movable cross-links.

3.3 Dynamics of Polyrotaxanes in Solution

One of the most interesting features of polyrotaxanes is the relative motion between the cyclic molecules and the backbone polymer chains with topological restriction. The cyclic molecules in polyrotaxanes can slide and rotate on the linear polymer chains, as shown in Figure 3.9. The unique intramolecular dynamics lead to the development of polyrotaxane-based nanomaterials: molecular shuttles in which the position of a ring molecule on a backbone can be switched by external stimuli,[29] polyrotaxane-ligand conjugates, which realize multivalent ligand–receptor interactions enhanced by the sliding/rotation motion,[30] and slide-ring materials with movable cross-links.[24]

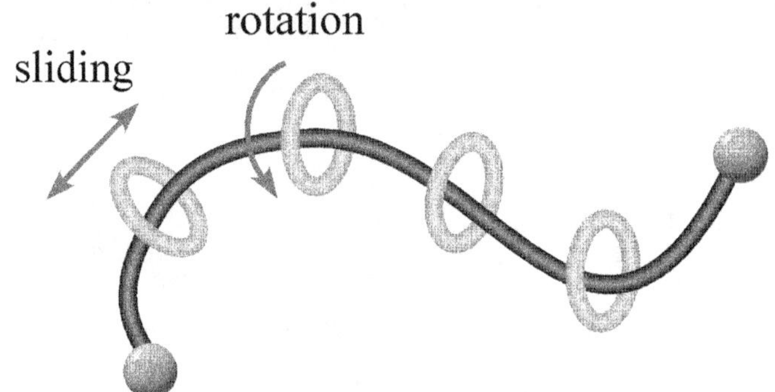

Figure 3.9   Schematic view of internal dynamics in a polyrotaxane.[4]
Reprinted from Polymer, 51, 959–967. Copyright (2010), with permission
from Elsevier.

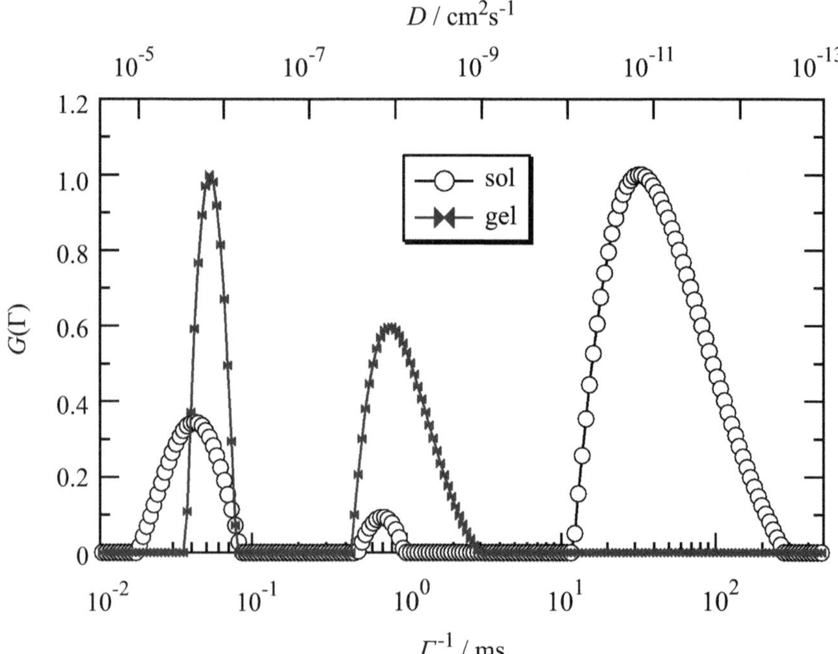

Figure 3.10   Decay time distribution functions $G(\Gamma)$ for polyrotaxane solutions and
slide-ring gels.[31]
Reproduced by permission of IOP Publishing.

Dynamic light scattering (DLS) experiments have observed three dynamic
modes for polyrotaxane solutions.[4,31,32] The decay time distributions $G(\Gamma)$
for semi-dilute polyrotaxane solutions and slide-ring gels formed by cross-
linking the polyrotaxanes are displayed in Figure 3.10. The slowest mode

with a relaxation time Γ^{-1} of about 30 ms is assigned to self-diffusion of the polyrotaxanes because it disappears in the slide-ring gels. The fastest mode with $\Gamma^{-1} = 0.04$ ms corresponds to the cooperative diffusion of the polyrotaxane networks. Between the two modes, a diffusive mode with $\Gamma^{-1} = 0.7$ ms is found both in the solutions and the gels. The intermediate mode is regarded as the sliding mode of the cyclic molecules in the polyrotaxane solutions and the slide-ring gels.

The segmental dynamics of polyrotaxanes in solution have been investigated by neutron spin echo (NSE) with much smaller length scales and shorter time scales than DLS ($0.01\,\text{Å}^{-1} < Q < 1\,\text{Å}^{-1}$, 0.01 ns $< t < 100$ ns).[4,33] NSE has been employed to study a wide range of dynamics in polymers such as reptation dynamics in polymer melts,[34] segmental motion in polymer melts and solutions,[34] and intramolecular dynamics in proteins.[35] Mayumi *et al.* have performed NSE measurements on polyrotaxanes in solution.[33] Figure 3.11 shows the Q dependence of the decay rate Γ for PEG and polyrotaxane solutions. In the case of the PEG solutions, Γ is proportional to Q^3, which is explained by the Zimm model for the segmental dynamics of flexible polymers in solutions. On the other hand, for the polyrotaxane solutions the Zimm dynamics are not observed in the experimental regime, and Γ is much smaller than that of the PEG solutions in the high-Q regime.

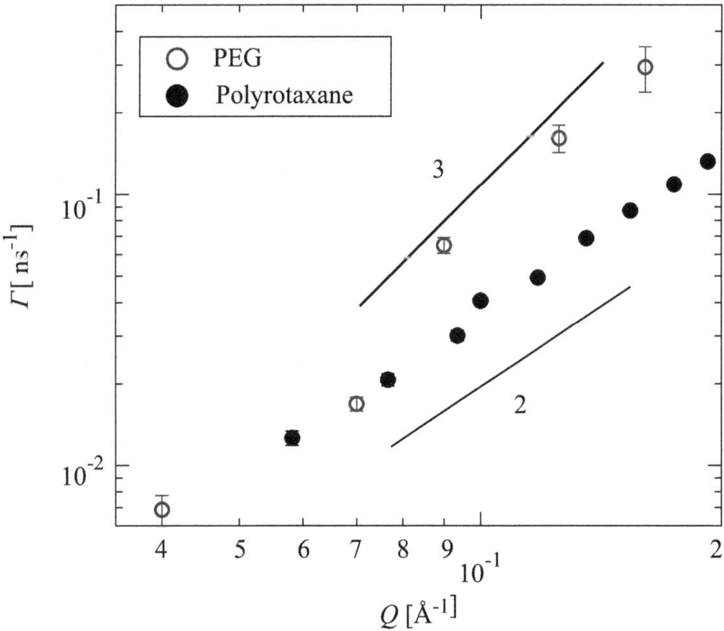

Figure 3.11   Double logarithmic plot of Γ as a function of Q for PEG solutions (open circles) and polyrotaxane (PR) solutions (closed circles).[33] Reprinted from Physica B, 404, 2600–2602. Copyright (2009), with permission from Elsevier.

The complexation of PEG with CDs slows down the local dynamics of the polymer chains.

References

1. M. J. Frampton and H. L. Anderson, *Angew. Chem., Int. Ed.*, 2007, **46**, 1028.
2. A. Harada, J. Li and M. Kamachi, *Nature*, 1993, **364**, 516.
3. T. Ooya and N. Yui, *J. Controlled Release*, 1999, **58**, 251.
4. K. Mayumi and K. Ito, *Polymer*, 2010, **51**, 959.
5. J. S. Higgins and H. C. Benoit, *Polymers and Neutron Scattering*, Clarendon Press, Oxford, 1994.
6. D. G. H. Ballard, G. D. Wignall and J. Schelten, *Eur. Polym. J.*, 1973, **9**, 965.
7. K. Mayumi, N. Osaka, H. Endo, H. Yokoyama, Y. Sakai, M. Shibayama and K. Ito, *Macromolecules*, 2008, **41**, 6580.
8. M. Rubinstein and R. H. Colby, *Polymer Physics*, Oxford University Press, 2003.
9. T. Yoshizaki and H. Yamakawa, *Macromolecules*, 1980, **13**, 1518.
10. T. Kume, J. Araki, Y. Sakai, K. Mayumi, M. Kidowaki, H. Yokoyama and K. Ito, *J. Phys.: Conf. Ser.*, 2009, **184**, 012018.
11. G. Fleury, C. Brochon, G. Schlatter, G. Bonnet, A. Lapp and G. Hadziioannou, *Soft Matter*, 2005, **1**, 378.
12. S. Yamada, Y. Sanada, A. Tamura, N. Yui and K. Sakurai, *Polym. J.*, 2015, **47**, 464.
13. K. Mayumi, H. Endo, N. Osaka, H. Yokoyama, M. Nagao, M. Shibayama and K. Ito, *Macromolecules*, 2009, **42**, 6327.
14. H. Endo, S. Miyazaki, K. Haraguchi and M. Shibayama, *Macromolecules*, 2008, **41**, 5406.
15. D. Richter, D. Schneiders, M. Monkenbusch, L. Willner, L. J. Fetters, J. S. Huang, M. Lin, K. Mortensen and B. Farago, *Macromolecules*, 1997, **30**, 1053.
16. H. Endo, D. Schwahn and H. Cölfen, *J. Chem. Phys.*, 2004, **120**, 9410.
17. H. Endo, K. Mayumi, N. Osaka, K. Ito and M. Shibayama, *Polym. J.*, 2011, **43**, 155.
18. C. Travelet, G. Schlatter, P. Hebraud, C. Brochon, A. Lapp, D. V. Anokhin, D. A. Ivanov, C. Gaillard and G. Hadziioannou, *Soft Matter*, 2008, **4**, 1855.
19. T. Karino, Y. Okumura, C. Zhao, M. Kidowaki, T. Kataoka, K. Ito and M. Shibayama, *Macromolecules*, 2006, **39**, 9435.
20. T. Kataoka, M. Kidowaki, C. Zhao, H. Minamikawa, T. Shimizu and K. Ito, *J. Phys. Chem. B*, 2006, **110**, 24377.
21. J. P. Gong, *Soft Matter*, 2010, **6**, 2583.
22. M. Shibayama, *Polym. J.*, 2011, **43**, 18.
23. A. Onuki, *J. Phys. II*, 1992, **2**, 45.
24. Y. Okumura and K. Ito, *Adv. Mater.*, 2001, **13**, 485.
25. K. Ito, *Polym. J.*, 2007, **39**, 489.

26. T. Karino, Y. Okumura, C. Zhao, T. Kataoka, K. Ito and M. Shibayama, *Macromolecules*, 2005, **38**, 6161.
27. E. Moses, T. Kume and T. Hashimoto, *Phys. Rev. Lett.*, 1994, **72**, 2037.
28. Y. Shinohara, K. Kayashima, Y. Okumura, C. Zhao, K. Ito and Y. Amemiya, *Macromolecules*, 2006, **39**, 7386.
29. S. A. Nepogodiev and J. F. Stoddart, *Chem. Rev.*, 1998, **98**, 1959.
30. N. Yui and T. Ooya, *Chem.-Eur. J.*, 2006, **12**, 6730.
31. C. Zhao, Y. Domon, Y. Okumura, S. Okabe, M. Shibayama and K. Ito, *J. Phys.: Condens. Matter*, 2005, **17**, S2841.
32. M. Shibayama, T. Karino, Y. Domon and K. Ito,*J. Appl. Crystallogr.*, 2007, **40**, s43.
33. K. Mayumi, M. Nagao, H. Endo, N. Osaka, M. Shibayama and K. Ito, *Physica B*, 2009, **404**, 2600.
34. D. Richter, M. Monkenbusch, A. Arbe and J. Colmenero, *Adv. Polym. Sci.*, 2005, **174**, 1.
35. R. Inoue, R. Biehl, T. Rosenkranz, J. Fitter, M. Monkenbusch, A. Radulescu, B. Farago and D. Richter, *Biophys. J.*, 2010, **99**, 2309.

Mechanical Properties of Slide-ring Materials

4.1   Overview of Slide-ring Materials

Since the discovery of cross-linking in natural rubber with sulfur in 1839 by Goodyear, the cross-linking of polymeric materials has become one of the most important topics in polymer science and technology.[1,2] Uncross-linked natural rubbers or elastomers inherently resemble liquids with regard to their flow behavior, although they exhibit viscoelastic properties based on the entanglement of polymer chains. If they are cross-linked, they behave like solids to maintain their shape against deformation at a temperature above the glass transition. Various unique mechanical properties of cross-linked polymeric materials have been investigated so far. Kuhn, and Mark and Erman assumed the affine deformation of cross-linking junction points and a Gaussian random chain between cross-links in order to explain the stress–strain curves of the cross-linked polymeric materials; this is called the fixed junction model.[2,3] Since then, this simple assumption has been re-considered by Guth and James,[4] Flory and Rehner,[5] Edwards and Freed,[6] and so on. For example, Guth and James took into account the fluctuation of the cross-linking junctions, which is referred to as the phantom network model.[4] Some experimental results obtained from small-angle neutron scattering (SANS) supported the affine deformation of the fixed junction model rather than the phantom network model.[7,8] However, the Tetra-poly(ethylene glycol) (Tetra-PEG) gel developed recently by Sakai *et al.* and Chung *et al.* shows quantitative agreement with the phantom network model.[9,10] This is because the gel synthesized by cross-linking two well-defined symmetrical tetra-arm polymers of the same size has a uniform and homogeneous structure with the same length of all network strands.

Monographs in Supramolecular Chemistry No. 15
Polyrotaxane and Slide-Ring Materials
By Kohzo Ito, Kazuaki Kato and Koichi Mayumi

Published by the Royal Society of Chemistry, www.rsc.org

The polymer entanglement is also one of the most important features for polymers. It behaves as the cross-link over a short time scale, but entangled chains are eventually released by reptation or one-dimensional diffusion along a polymer chain. Accordingly, the reptation of uncross-linked chains relaxes the rubber elasticity of the entangled polymer system completely at the longest relaxation time. It is well known that the entanglement effect was described effectively by the tube or reptation model developed by de Gennes, and Edwards and Doi in the 1970s, in which each chain is assumed to move along a tube due to constraints imposed by surrounding chains.[11,12] The tube model has enabled us to understand the relationship between molecular structure and linear rheology in an entangled system. On the other hand, Ball *et al.* also proposed a slip-link model in which the chain can slip at a trapped entanglement.[13] The slip-link model is widely used in computer simulations of the viscoelastic behavior for uncross- and cross-linked polymeric materials.

When cross-linked polymeric materials are immersed in a good solvent, they absorb the liquid solvent until the swelling force associated with the mixing entropy between the chains and the solvent balances the elastic force of the chains between junction points. These cross-linked polymeric systems containing the solvent are called chemical gels. The swelling behavior of the chemical gels was explained in detail by Flory and Rehner.[14] On the other hand, Tanaka discovered the volume phase transition of the chemical gels, in which the swelling and shrinking behaviors exhibited discontinuous profiles with hysteresis.[15,16] This novel discovery regarding the cross-linked polymeric materials has attracted considerable interest from researchers in the field of polymer science. As a result, some interesting aspects of the chemical gels have been discovered one after the other; these aspects include the kinetics of the volume phase transition by Tanaka and Fillmore,[17] the frozen or fixed inhomogeneous structure of the chemical gels by Shibayama,[18] and the abnormally small friction behavior of gels by Gong.[19] In addition, various new types of gels have been developed thus far. For example, Gong *et al.* developed a double network gel having a high modulus up to the sub-mega Pascal range with a failure compressive stress as high as 20 MPa.[20] The double network gel has both soft and hard components so as to avoid fracture, as in some biomaterials.[21] Furthermore, Yoshida incorporated a dissipative system into gels to realize a self-oscillating gel device,[22] and also reported comb-type grafted hydrogels showing a rapid de-swelling response to temperature changes.[23]

All gels are classified into two categories: chemical gels and physical gels.[24] Physical gels have noncovalent cross-linking junctions from ionic interactions, hydrophobic interactions, hydrogen bonding, microcrystal formation, helix formation, and so on. In general, these noncovalent cross-links are not as strong as the covalent cross-links in chemical gels, and the physical gels show a sol–gel transition response to temperature, pH, and solvent. The mechanical behavior of the physical gels is complicated because the recombination of cross-linking points occurs on deformation; hence, the affine deformation is not validated in the physical gel. The recombination causes hysteresis in the stress–strain curve of the physical gel, which means that it cannot regain its original shape from the deformed one as quickly as the chemical gel. Haraguchi and Takeshita have developed a novel type of gel called the nano-composite gel, which has clay as the cross-linking junction.[25] Polymer chains of *N*-isopropyl acrylamide strongly absorb onto the clay surface, thus bridging different clays. Surprisingly, the nano-composite gel shows high stretchability up to 10 times its original length. This suggests that the structures of the cross-linking junctions play an important role in determining the mechanical properties of polymeric materials.

Another recent approach for the polymer network structure has been developed using polyrotaxane in the supramolecular chemistry. The first report of a physical gel based on the polyrotaxane architecture was done by Harada *et al.*[26] When α-cyclodextrins were mixed with long poly(ethylene glycol) chains at a high concentration in water, the sol–gel transition occurred due to hydrogen bonding between the α-cyclodextrins threaded on the poly(ethylene glycol) chains in different pseudopolyrotaxanes. In addition, Yui and coworkers formed some hydrogels using biodegradable polyrotaxane as the cross-linker for use in regenerative medicine as will be mentioned in more detail in Chapter 8.[27] The biodegradable polyrotaxane has a hydrolysis part, namely, an ester bond between a bulky end group and the polymer axis. Consequently, the erosion time of the biodegradable hydrogel strongly depends on its polyrotaxane content. Furthermore, Takata synthesized recyclable cross-linked polyrotaxane gels: topologically networked polyrotaxane capable of undergoing reversible assembly and disassembly based on the concept of dynamic covalent bond chemistry.[28] This work described cross-linked poly(crown ether)s with dumb-bell-shaped axle molecules, which showed a reversible cleavage of the disulfide bond. As a result, a novel reversible cross-linking/decross-linking system that could recycle networked polymeric materials was realized.

Okumura and Ito developed another cross-linking structure based on the polyrotaxane architecture, as mentioned in Chapter 1.[29] They prepared polyrotaxane sparsely containing α-cyclodextrins and subsequently cross-linked α-cyclodextrins on different polyrotaxanes as shown in Figure 1.2. As a result, the cross-linking junctions of figure-of-eight shapes were not fixed at the poly(ethylene glycol) chains and can move freely in the polymer network. They refer to this new cross-linked polymer network as a *topological gel*, or a *slide-ring material* in a wider sense. Such a polymeric material with freely movable cross-links was theoretically considered as a sliding gel by de Gennes in 1999.[30] In addition, the historical significance of the slide-ring materials or gels was reviewed compared to the slip-link model.[31] It is noteworthy that the concept of freely movable cross-links is not limited within the slide-ring gel containing some solvents but can also be applied to slide-ring polymeric materials without solvents. As mentioned in the following section, it is clear that the freely movable cross-link drastically changes the mechanical properties of polymeric materials. This may bring about a paradigm shift in cross-linked polymeric materials since the cross-linking was first discovered by Goodyear.

4.2 Pulley Effect and the Stress–Strain Curve

As mentioned in the previous section, all gels are categorized into physical or chemical gels according to their network junctions. The physical gel has noncovalent cross-links so that it can be easily liquefied at high temperature or in a good solvent. On the other hand, the chemical gel with covalent cross-links is more stable than the physical gel, and therefore it is used in many applications such as in soft contact lenses, and so on. However, the chemical gel generally has a large inhomogeneous structure due to the gelation process, which considerably reduces the mechanical strength. In the chemical gel, long polymer chains are divided into shorter pieces of different lengths by fixed cross-links. As a result, the tensile stress is concentrated on the shortest chains, thereby breaking down the chemical gel easily.

In the topological or slide-ring gel, on the other hand, the figure-eight-shaped cross-links can pass along the polymer chains freely in order to equalize the tension of the threading polymer chains in a manner similar to pulleys; this is called the pulley effect.[29] As a result, the slide-ring gel shows a high stretchability of up to 24 times in length, and a huge volume change of up to 24 000 times in weight. Furthermore, the pulley effect yields various unique properties of the slide-ring gel that are different from the chemical gel.

The mechanical properties of the slide-ring gel are quite different from those of conventional physical and chemical gels.[32,33] The physical gel shows a J-shaped stress–strain curve with large hysteresis, as shown in Figure 4.1. The large hysteresis is caused by recombination among

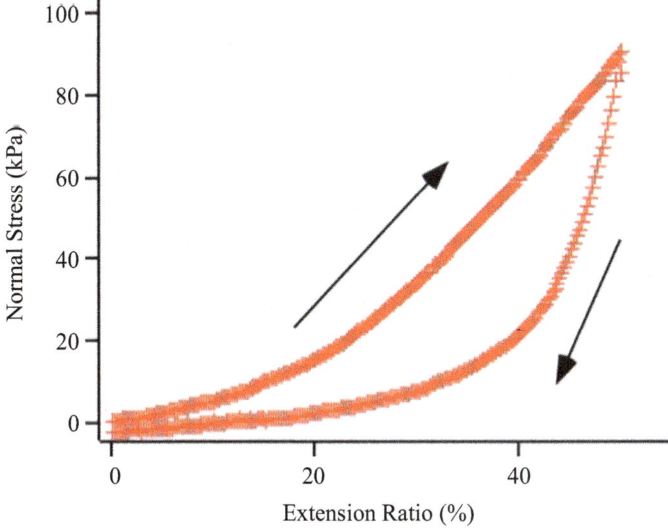

Figure 4.1 Stress–strain curve of konjak as a typical physical gel. The physical gel shows a concave stress–strain curve with large hysteresis.[32]
Reprinted with permission from *Polym. J.* Copyright (2007) Macmillan Publishers Ltd.

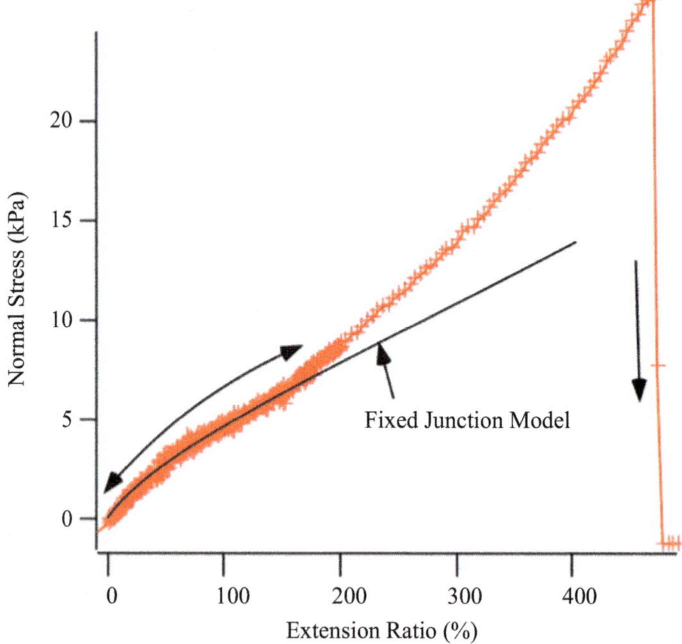

Figure 4.2   Stress–strain curve of poly(2-acrylamido-2-methylpropane sulfonic acid) (PAMPS) gel. The chemical gel shows an S-shaped stress–strain curve without hysteresis. The stress–strain behavior in low extension ratio is well described by the fixed junction model.[32]
Reprinted with permission from *Polym. J.* Copyright (2007) Macmillan Publishers Ltd.

noncovalent cross-links in the polymer network on deformation. On the other hand, the chemical gel shows no hysteresis since it exhibits stable covalent cross-links without recombination on deformation, as shown in Figure 4.2. In addition, the chemical gel shows an S-shaped stress–strain curve, similar to that of cross-linked natural rubber, which is well explained by the three chain model or the fixed junction model. In the fixed junction model, assuming the affine deformation of the junction points and the additivity of the individual conformational entropies of the three Gaussian chains, we can obtain the dependence of the normal stress σ on the extension ratio λ for a uniaxial deformation using the following well-known expression:[1,2]

$$\sigma = \nu k_{\mathrm{B}} T (\lambda - \lambda^{-2}), \tag{4.1}$$

where ν is the number density of network strands, k_{B} is the Boltzmann constant, and T is the absolute temperature. The stress–strain curve of the chemical gel shows a concave-down profile in the low extension region, which is well-fitted by eqn (4.1), while it shows a concave-up profile with deviation from eqn (4.1) in the high-extension region because of

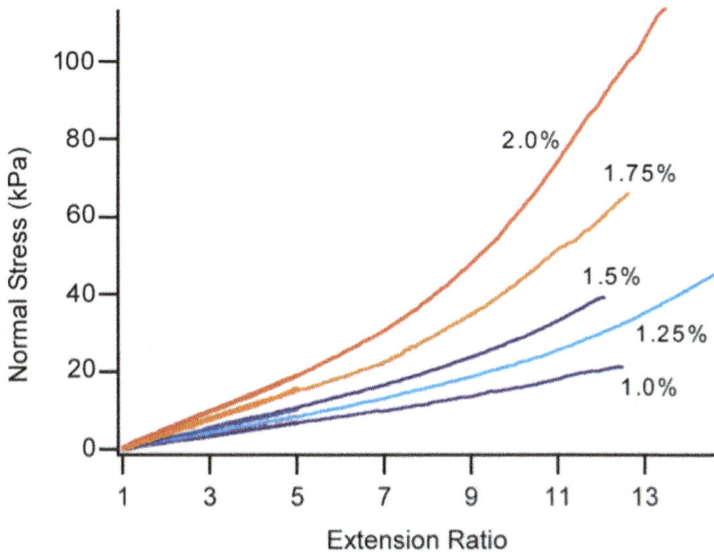

Figure 4.3 Stress–strain curve of the slide-ring gel with different concentrations of cyanuric chloride as a cross-linker. The slide-ring gel shows a J-shaped stress–strain curve without hysteresis.[32]
Reprinted with permission from *Polym. J.* Copyright (2007) Macmillan Publishers Ltd.

non-Gaussian behavior or the stretching effect. As a result, the chemical gel shows the S-shaped stress–strain curve in Figure 4.2.

The slide-ring gel exhibited a J-shaped curve different from the chemical gel; furthermore, it showed no hysteresis loop, in contrast to the physical gel, as shown in Figure 4.3.[32] This means that the slide-ring gel is soft compared to the chemical gel. de Gennes proposed a model of a sliding gel, which has movable cross-links, and discussed the mechanical properties theoretically.[30] Contrary to native expectation, the sliding gel should not be particularly soft at low stresses because all the stored length is used to swell the gel. Therefore, the theory could not explain the softness of the slide-ring gel.

Ito considered a modified three chain model or free junction model corresponding to the pulley effect, in which three Gaussian chains were assumed to be able to slide toward each other in the following manner.[32] Let us consider that three chains in the x-, y-, and z-directions have N_x, N_y, and N_z segments, respectively, and displacement lengths R_x, R_y, and R_z in that order, where the total chain length $3N$ is constant:

$$N_x + N_y + N_z = 3N. \tag{4.2}$$

Then, the number of states W is given by:

$$W = \left(\frac{3}{2\pi b^2}\right)^{3/2} \frac{1}{\sqrt{N_x N_y N_z}} \exp\left[-\frac{3}{2b^2}\left(\frac{R_x^2}{N_x} + \frac{R_y^2}{N_y} + \frac{R_z^2}{N_z}\right)\right], \tag{4.3}$$

where b is the segment length. If the gel is deformed uniaxially in the z-direction, R_x, R_y, and R_z are related to λ as $R_x = R_y = R_0/\sqrt{\lambda}$ and $R_z = \lambda R_0$, where $R_0 = \sqrt{N}b$ is the displacement at equilibrium. Then, the entropy S is written as:

$$S(N_x, N_y, N_z\,;\,\lambda) = -\frac{k_B}{2}\left[\ln(N_xN_yN_z) + 3N\left(\frac{1}{\lambda N_x} + \frac{1}{\lambda N_y} + \frac{\lambda^2}{N_z}\right)\right]$$
$$+\frac{3k_B}{2}\ln\left(\frac{3}{2\pi b^2}\right). \tag{4.4}$$

Here, we maximize the entropy with respect to N_x, N_y, and N_z as $\partial S/\partial N_x = \partial S/\partial N_y = 0$, which leads to $N_x = N_y$, and $N_z(\lambda)$ as a function of λ obtained by solving the following equation:

$$\frac{1}{N_z} - \frac{\lambda^2 N}{N_z^2} = \frac{2}{3N - N_z} - \frac{4N}{\lambda(3N - N_z)^2}. \tag{4.5}$$

Then, the total free energy F is given by:

$$F(\lambda) = \frac{n}{2}k_\mathrm{B}T\left[\ln\frac{N_z(3N - N_z)^2}{4} + N\left(\frac{\lambda^2}{N_z} + \frac{4}{\lambda(3N - N_z)}\right)\right]. \tag{4.6}$$

As a result, the normal stress σ can be expressed as $\sigma(\lambda) = \partial F(\lambda)/\partial(\lambda V)$, where n and V are the total number of network strands between the cross-links and the total volume of the gel, respectively.

Figure 4.4 shows the comparison between the fixed and free junction models with respect to the stress–strain curve in a low extension region. The fixed junction model shows the concave-down stress–strain curve, as mentioned earlier, while the free junction model exhibits the concave-up curve, which agrees qualitatively with the J-shaped experimental results of the slide-ring gel, as shown in Figure 4.3. The difference is also observed in the compression region (Figure 4.5). The free junction model indicates that the slide-ring gel shows a significantly small amount of normal stress at a low compression ratio, and the normal stress increases drastically with the compression ratio increases over $\lambda \approx 0.4$. On the other hand, the fixed junction model shows that the normal stress increases gradually with compression. It is noted that this difference is not limited only to the gels, but should also be observed in the cross-linked polymeric materials without solvents. Consequently, the pulley effect arising from freely movable cross-links provides significantly different mechanical properties from the fixed cross-linking junctions.

Next, we consider the biaxial deformation of the cross-linked polymeric materials and balloon inflation. In the fixed junction model, the dependence of the surface tension γ on the extension ratio λ is given by: [32]

$$\gamma = \frac{n}{A}k_\mathrm{B}T(1 - \lambda^{-6}), \tag{4.7}$$

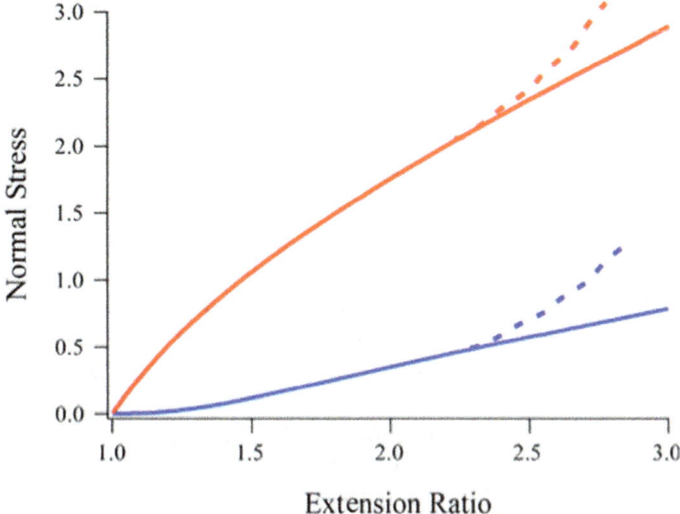

Extension Ratio

Figure 4.4 Comparison between the fixed (upper, red) and free (lower, blue) junction models with regards to the stress–strain curves on uniaxially stretched deformation. The doted curves reflect the deviation from the Gaussian chains.[32]
Reprinted with permission from *Polym. J.* Copyright (2007) Macmillan Publishers Ltd.

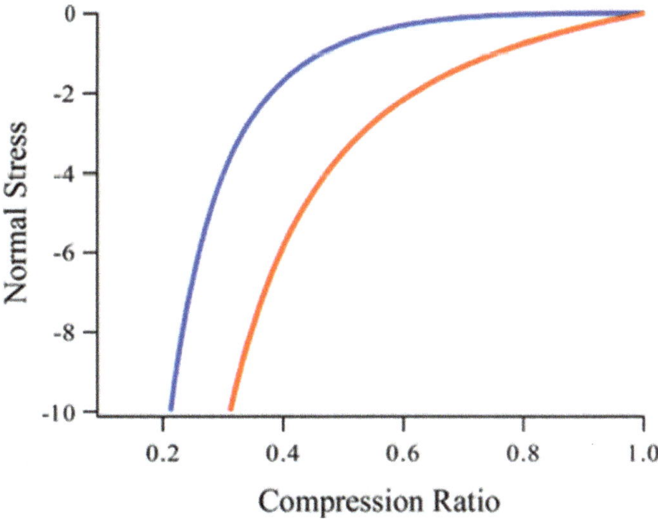

Compression Ratio

Figure 4.5 Comparison between the fixed (lower, red) and free (upper, blue) junction models with regards to the stress–strain curves on uniaxially compressed deformation.[32]
Reprinted with permission from *Polym. J.* Copyright (2007) Macmillan Publishers Ltd.

where A denotes the unstretched cross-section area in the initial state before inflation. Then, the pressure p in a spherical balloon is given by:

$$p = \frac{3n}{V_0} k_\text{B} T \left(\frac{1}{\lambda} - \frac{1}{\lambda^7} \right), \tag{4.8}$$

where V_0 is the volume of the spherical balloon in the initial state before inflation. This has a maximum at $\lambda = 7^{1/6} \approx 1.383$, as shown in Figure 4.6, which was verified experimentally.[34] However, similar to the stress–strain curve, as mentioned earlier, the actual p–λ profile turns upwards, thereby deviating from eqn (4.8) in the high-extension region over $\lambda \approx 2.5$ because of the non-Gaussian behavior or stretching effect. As a result, a cylindrical balloon exhibits a bistable state, called *elastic instability*, in which the swollen and shrunken parts coexist in the balloon. This elastic instability is also observed in the generation of bubbles during foam production, and the development of aneurysms in arteries in the human body.[35]

The free junction model provides a significantly different p–λ profile. If the gel is deformed biaxially in the x and y directions, R_x, R_y, and R_z are related to the extension ratio λ as $R_x = R_y = \lambda R_0$ and $R_z = R_0/\lambda^2$. Therefore, the entropy is written as:

$$S(N_x, N_y, N_z\,;\,\lambda) = -\frac{k_\text{B}}{2}\left[\ln(N_x N_y N_z) + 3N\left(\frac{\lambda^2}{N_x} + \frac{\lambda^2}{N_y} + \frac{1}{\lambda^4 N_z} \right) \right]$$
$$+ \frac{3k_\text{B}}{2}\ln\left(\frac{3}{2\pi b^2} \right). \tag{4.9}$$

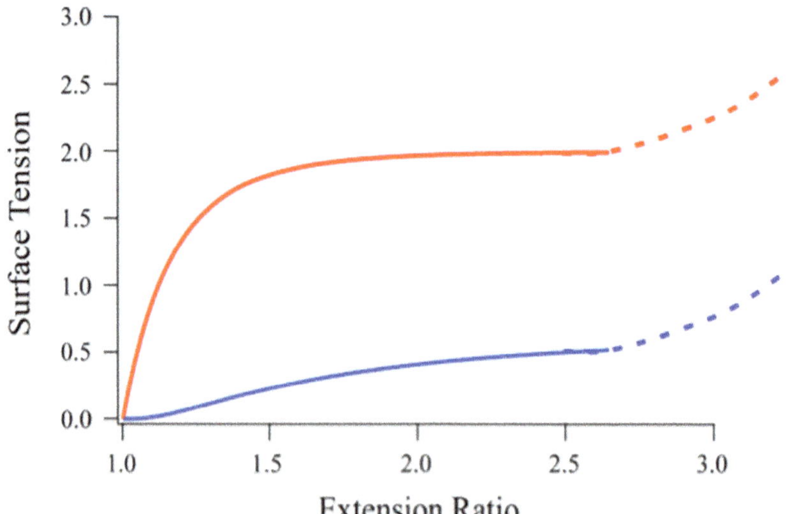

Figure 4.6 Comparison between the fixed (upper, red) and free (lower, blue) junction models with regards to surface tension on biaxial deformation. The doted curves reflect the deviation from the Gaussian chains.[32]
Reprinted with permission from *Polym. J.* Copyright (2007) Macmillan Publishers Ltd.

By maximizing the entropy with respect to N_x, N_y, and N_z as $\partial S/\partial N_x = \partial S/\partial N_y = 0$, which leads to $N_x = N_y$, and $N_z(\lambda)$, we have $N_x = N_y$, and $N_z(\lambda)$ as a function of λ obtained by solving the following equation:

$$\frac{1}{N_z} - \frac{N}{\lambda^4 N_z{}^2} = \frac{2}{3N - N_z} - \frac{4\lambda^2 N}{(3N - N_z)^2}. \tag{4.10}$$

Then, the total free energy is given by:

$$F(\lambda) = \frac{n}{2} k_{\mathrm{B}} T \left[\ln \frac{N_z(3N - N_z)^2}{4} + N \left(\frac{1}{\lambda^4 N_z} + \frac{4\lambda^2}{3N - N_z} \right) \right]. \tag{4.11}$$

In addition, the surface tension γ and balloon pressure p are calculated as:

$$\gamma = \frac{1}{A} \frac{\partial F(\lambda)}{\partial \lambda^2} \tag{4.12}$$

$$p = \frac{2}{3\lambda^2 V_0} \frac{\partial F(\lambda)}{\partial \lambda}. \tag{4.13}$$

Figure 4.6 and Figure 4.7 show the comparison between the free and fixed junction models with regard to the dependence of surface tension γ and p, respectively, on λ. As λ increases, p in the free junction model has a maximum at $\lambda \approx 2.132$, and then decreases slightly and gradually; however, the actual p–λ profile turns upwards in the high-extension region over $\lambda \approx 2.5$

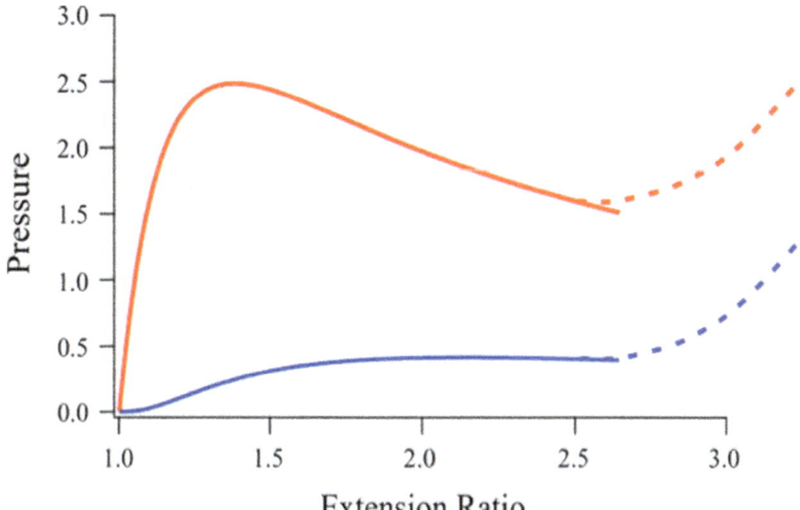

Figure 4.7 Comparisons between the fixed (upper, red) and free (lower, blue) junction models with regards to pressure on biaxial deformation. The doted curves reflect the deviation from the Gaussian chains.[32]
Reprinted with permission from *Polym. J.* Copyright (2007) Macmillan Publishers Ltd.

because of the non-Gaussian behavior or stretching effect, as mentioned earlier. It is thus clear that the free junction model provides no elastic instability that may cause aneurysms. Hence, the slide-ring materials with freely movable cross-links are more favorable for use in artificial arteries than fixed cross-linking materials.

Many biomaterials, such as mammalian skin, vessels, and tissues, show J-shaped stress–strain curves, which usually provide toughness and no elastic instabilities, among other advantages.[36,37] The J-shaped stress–strain curve yields toughness because its low shear modulus drastically reduces the energy released in the fracture, driving crack propagation. In addition, the material becomes stiffer as the extension ratio approaches the fracture point. The slide-ring materials show J-shaped stress–strain curves that are similar to the curves shown by biomaterials such as mammalian skin, vessels, and tissues. This indicates that the slide-ring materials are suitable for use as substitutes for various types of biomaterials. If artificial arteries are made of the slide-ring materials, they may fit better with the native ones than fixed cross-linked materials.

4.3   Entropy of the Rings and the Sliding Transition

Polyrotaxane consists of threaded cyclic molecules (rings) and a backbone axis polymer (string). This means that polyrotaxane has two kinds of entropy unlike usual polymers: the distribution or translational entropy of the rings, and the conformational entropy of the backbone string. The axis string can take various conformations, from coiled to rod-like, and almost regardless of the distribution of the rings. On the other hand, the threaded rings can have various distributions along the backbone string, from random to aggregated, and almost regardless of the string conformation. There is certainly some correlation between the two kinds of entropy, such that the string tends to take a stretched conformation in the part where the rings form aggregates. However, if the inclusion ratio is not so high, the correlation does not seem to affect the string conformation so significantly. This means that even if the conformation of polyrotaxane is observed in solution, it is not easy to detect the influence of the entropy of the rings on the polyrotaxane conformation. Actually, Mayumi and Ito succeeded in the detection of a slight influence of the entropy of the rings on the polyrotaxane conformation as an increase of the persistence length of the axis string, poly(ethylene glycol), as mentioned in Chapter 3.[38] Consequently, the two kinds of entropy in polyrotaxane are almost independent of each other.

The independence, however, is changed drastically by cross-linking polyrotaxane to form slide-ring materials.[39] Experimental results show that a number of cyclodextrins remain uncross-linked in the slide-ring gel. These cyclodextrins can slide along an axis polymer in polyrotaxanes in solution, resulting in the distribution entropy of the rings. Figure 4.8 schematically shows the molecular origin of the elasticity in the slide-ring gel. When the slide-ring gel is stretched in the horizontal axis, polymer chains are

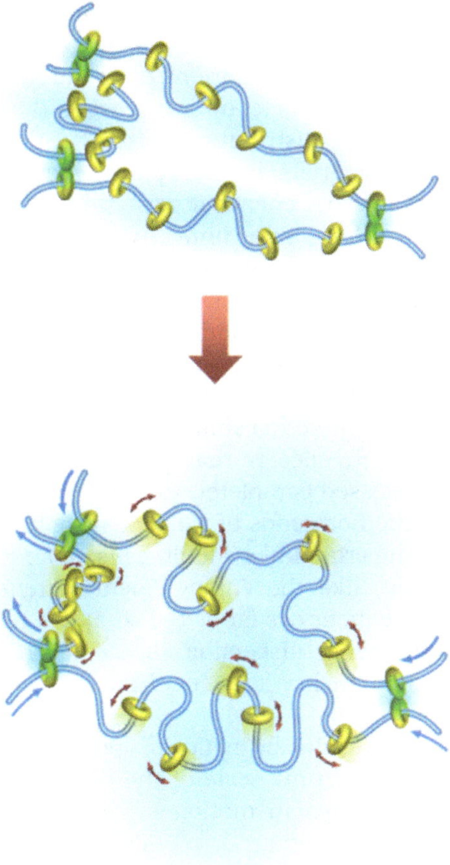

Figure 4.8 The relaxation of the rubber state to the sliding one by the sliding of axis polymer chains at the cross-linking junctions. The polymer sliding relaxes the conformation anisotropy of the axis polymer chains. Then, free cyclic molecules form heterogeneous density distribution since they cannot pass through cross-links consisting of other rings with the same size.[39]
Reprinted with permission from *Polym. J.* Copyright (2011) Macmillan Publishers Ltd.

deformed to anisotropic conformation over short time scales. As time passes, the axis polymer relaxes the polymer deformation by sliding at the cross-links, which is called the pulley effect. However, free rings cannot pass through the cross-links consisting of other rings of the same size, while axis polymer strings can easily do so. As a result, the axis polymer sliding forms heterogeneous linear density distribution of free rings, while polymer chains relax the conformation anisotropy to the isotropic form. The heterogeneous distribution reduces the distribution entropy of the free molecules drastically, which leads to the entropic elasticity from the entropy of the rings.

This is completely different from the entropic elasticity or rubber elasticity due to the polymer conformation. Since the entropy reduction shifts from string conformation to ring distribution, the slide-ring asymmetry of the rings and strings yields dynamic coupling between the conformational entropy of the strings and the distribution entropy of the rings by the pulley effect in the slide-ring materials, including the slide-ring gels.

It is well known that usual chemical gels and rubbers have finite equilibrium moduli (in the long time limit), which arise from the entropic elasticity of the polymer conformation between fixed cross-links. On the other hand, entangled polymer chains have no finite equilibrium modulus, like liquids, since the release or reptation of entangled polymers eventually relaxes the chain deformation responsible for the entropic elasticity. The slide-ring materials have a sliding motion of the backbone polymer chains at the cross-linking junction similar to the entangled polymer system. However, the largest difference is that polymer chains in the slide-ring materials cannot be released completely from the cross-links because there are bulk stoppers at the both ends of axis polymers. The sliding motion of the axis polymers at the cross-links should cause a frequency dispersion in the viscoelastic profile, like the entanglement system, but ought to be stopped at a time or frequency as far as the complete release is impossible. Consequently, the frequency dispersion should reach to a plateau (finite equilibrium modulus) in the low-frequency limit in the viscoelastic profile of slide-ring materials.

Recent experimental results by neutron spin echo and dynamic light scattering indicate that the time scale of the cyclodextrin sliding is much slower than the micro-Brownian motion of the poly(ethylene glycol) segment, as mentioned in Chapter 3.[38,40] This means that the polymer deformation should occur much faster than both the sliding relaxation of the axis polymer chain by the pulley effect and the sliding motion of the free rings. Accordingly, the slide-ring materials have two plateaus in different frequency ranges: the rubber plateau in the high-frequency range, where axis polymer strings do not slide at all, and the other one (finite equilibrium moduli) in the low-frequency limit, where the slide of strings is stopped after having occurred to some extent. In the high-frequency region, the slide-ring materials behave as usual chemical gels or cross-linked rubbers because the cross-linking cannot move. Then, the elasticity is due to the conformational entropy of polyrotaxane between cross-liking junctions or the micro-Brownian motion of polyrotaxane. On the other hand, the pulley effect works well in the low-frequency limit, where the entropy of the rings plays a dominant role in the finite equilibrium moduli. From these results, we suggest the dynamics of the slide-ring gel as schematically shown in Figure 4.9. Usual cross-linked amorphous polymers show glass and rubber states, which are divided by the glass transition. The slide-ring gel ought to have another state, what we have termed a *"sliding state"*, where axis polymer chains and cyclic molecules are sliding actively. The sliding state appears in higher temperature or lower frequency regions than the rubber state since the sliding motion of the free cyclodextrins is slower than the

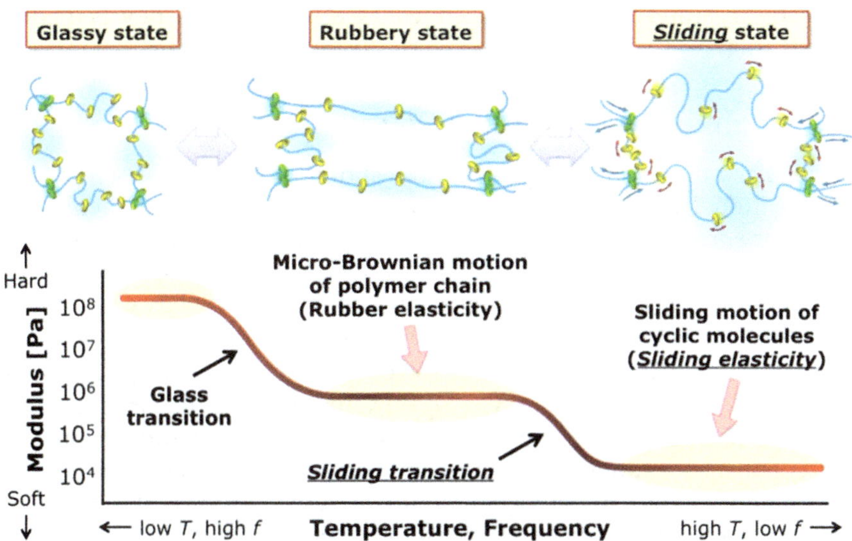

Figure 4.9 Mechanical dynamics of the slide-ring materials. Usual cross-linked amorphous polymers show the glass transition between glass and rubber states, while the slide-ring materials have the sliding transition and sliding state, where backbone polymer strings and cyclic molecules are sliding actively. The sliding state should appear in higher temperature or lower frequency regions than the rubber state.[39] Reprinted with permission from *Polym. J.* Copyright (2011) Macmillan Publishers Ltd.

micro-Brownian motion of the polymer segment. These two states are separated by the sliding transition, similar to the glass transition between the glass and rubber states. The sliding transition time reflects when the axis polymer chains and cyclic molecules start sliding.

If the distribution entropy of the rings dominates the sliding elasticity, the modulus should increase almost linearly with how many rings are included in the polyrotaxane, as discussed in Section 4.4. This means that the sliding elasticity is similar to a one-dimensional air spring, in which air molecules are confined in a one-dimensional tube, as mentioned in Section 2.3.[41] In the slide-ring gel, the sliding motion of free rings threaded onto a polymer chain is confined within movable cross-linking junctions. If the sliding elasticity is larger than the rubber one, the sliding transition and state should disappear since the polymer chain can deform easily rather than form the heterogeneous density distribution of free cyclic molecules by the polymer sliding at the cross-linking point. The entropy of the rings can also explain the qualitative deviation from de Gennes' prediction on the sliding gel. As mentioned in Section 4.2, de Gennes predicted that the sliding gel with movable cross-links swells spontaneously in the solvent since all the stored strand is used to swell the gel.[30] However, such a spontaneous swelling does not seem to be observed in the slide-ring gel actually. This may be because the entropy of the free rings disturbs the sliding of the stored strands in the slide-ring gel. This behavior was treated theoretically.[42]

4.4 Evaluation of Sliding Elasticity

The elasticity of the slide-ring gels in the sliding state originates from the conformational entropy of the liner polymers and the one-dimensional alignment entropy of the ring molecules. Mayumi *et al.* proposed a simple model to estimate the sliding elasticity based on the free junction model described in Section 4.2.[43] Consider a three chain model where each strand in the i direction has N_i segments ($i = x, y, z$). As explained in Section 4.2, the segment number N_i changes with deformation $\vec{\lambda} = (\lambda_x, \lambda_y, \lambda_z)$ under the restriction condition $N_x + N_y + N_z = 3N$ because the polymer strands can move through the movable crosslinks. The conformational entropy of the polymer strands S_P is given by eqn (4.4). In the case of the free junction model, the Young's modulus is zero, as shown in Figure 4.4. This is because the polymer chains can pass freely through the cross-links, and the chain conformation does not change under small strains. Hence, we have to consider another origin of the entropic elasticity in slide-ring materials: the distribution entropy of the rings. Here, we introduce ring molecules on the polymer chains (Figure 4.10). The ring molecules can slide on the polymer strands, which results in the one-dimensional alignment entropy of the rings. The entropy of the ring molecules S_R is given as follows:

$$S_R = k_B \ln \prod_i^{x,y,z} \binom{N_i}{n},\qquad(4.14)$$

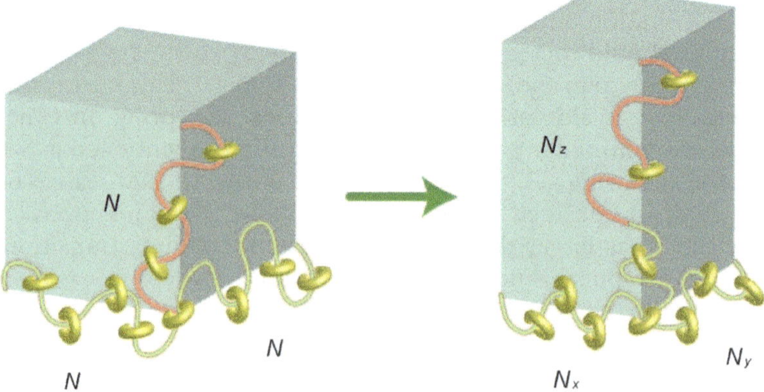

Figure 4.10 Three chain model for the polymer network with movable cross-links and uncross-linked cyclic molecules on polymer chains.
Reproduced from ref. 43 by permission of The Royal Society of Chemistry.

where n is the number of the rings on each polymer strand. To maximize the total entropy $S_{\text{total}} = S_P + S_R$, $N_i(\vec{\lambda})$ should satisfy the following condition:

$$\frac{\partial S_{\text{total}}}{\partial N_x} = \frac{\partial S_{\text{total}}}{\partial N_y} = \frac{\partial S_{\text{total}}}{\partial N_z}. \tag{4.15}$$

From eqn (4.2), (4.4), (4.14), and (4.15), the nominal stress σ_{SR} and Young's modulus E_{SR} of the sliding network with the ring molecules for uniaxial deformation $\vec{\lambda} = (\lambda^{-1/2}, \lambda^{-1/2}, \lambda)$ are described in simple forms:

$$\sigma_{\text{SR}} = -\frac{T}{V}\frac{\partial S_{\text{total}}}{\partial \lambda} = \nu k_{\text{B}} T \left[\frac{N}{N_z}\lambda - \frac{N}{N_x}\frac{1}{\lambda^2}\right] \tag{4.16}$$

$$E_{\text{SR}} = 3\nu k_{\text{B}} T - 9\nu k_{\text{B}} T \frac{1 - f_{\text{eff}}}{(2N - 3)f_{\text{eff}} + 3} \equiv E_{\text{rubber}} - \Delta E, \tag{4.17}$$

where $f_{\text{eff}} = n/N$ is the filling ratio of the ring molecules on the polymer chains between cross-links. E_{rubber} and ΔE are the elastic modulus for the conventional rubber elasticity theory and the gap between E_{rubber} and E_{SR}, respectively. ΔE corresponds to the effect of the sliding of the polymer chains on the elastic modulus. When there are no rings on the chains $(f_{\text{eff}} = 0)$, E_{SR} becomes zero, which is consistent with the free junction model without rings. If f_{eff} is much smaller than 1, E_{SR} is proportional to the filling ratio f_{eff}:

$$E_{\text{SR}} = 2\nu k_{\text{B}} T N f_{\text{eff}} \quad (f_{\text{eff}} \ll 1). \tag{4.18}$$

This can be seen as an analogy to the elasticity of the air spring.

By using the model, we can predict the ν dependence of the elastic modulus. ν is a function of the segment number N between cross-links and the segment length b:

$$\nu = \frac{3}{(N^{1/2}b)^3}. \tag{4.19}$$

f_{eff} decreases with the increase of ν since two ring molecules are used for one movable crosslink:

$$f_{\text{eff}} = f - \frac{2}{3N}. \tag{4.20}$$

Here, f is the filling ratio of uncross-linked polyrotaxanes. Figure 4.11 shows the ν dependence of E_{SR} calculated from eqn (4.17), (4.19), and (4.20) ($b = 1\,\text{nm}$, $f = 0.25$, and $T = 298$ K). At the sliding transition where the polymers and ring molecules start sliding, the elastic modulus changes from E_{rubber} to E_{SR}. As shown in Figure 4.11, the gap between E_{rubber} and E_{SR} increases with ν, which is consistent qualitatively with the experimental results.[44] A quantitative comparison between the theory and experimental results is in progress.

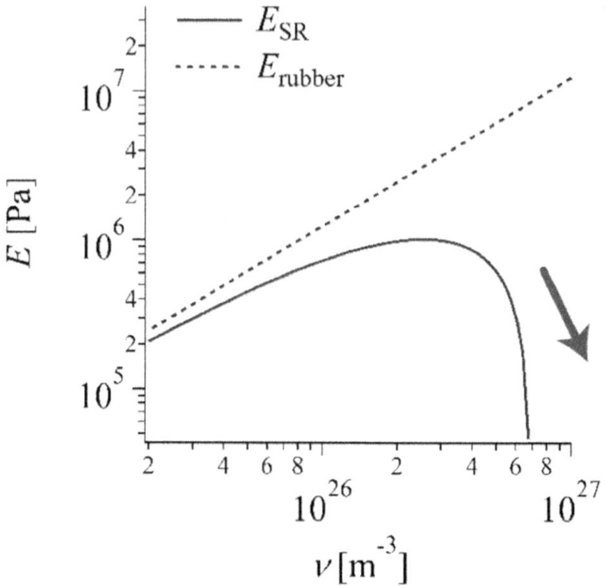

Figure 4.11   Double logarithmic plot of calculated Young's moduli E for conventional chemical gels (dashed line), and SR gels (solid line) as a function of the number density of network strands, ν.
Reproduced from ref. 43 by permission of The Royal Society of Chemistry.

4.5   Viscoelastic Profile of Slide-ring Materials

According to the time scale of the chain sliding through the cross-link, a characteristic relaxation is observed as long as the sliding elasticity is smaller than the rubber one. Figure 4.12 shows the viscoelastic data of a series of slide-ring gels that are prepared by the cross-linking of a poly-butadiene-based polyrotaxane with different cross-linking concentrations (see Section 7.4.2.1).[44,45] The coverage of the precursor polyrotaxane is 30%, and the molecular weights (M_n) of the backbone polybutadiene and polyrotaxane are 4000 and 19 000, respectively. Two plateaus of storage modulus are observed at high- and low-frequency limits, and the relaxation changes the modulus more than 10 times. The modulus at the high-frequency limit is 100–400 kPa, typical values for polymer gels. From the modulus at the high-frequency limit E_r, cross-linking density ν and averaged molecular weight between cross-links M_x can be estimated by using the following equation, which is known as rubber elastic theory:[1,2]

$$E_r = 3\nu kT = \frac{3\rho RT}{M_x},\qquad(4.21)$$

where ρ is the density of the elastic body, and R is the gas constant. M_x is calculated to be 3000–9000 g mol^{-1}. This value indicates the possible picture of the network structure, where several cross-links exist in a single

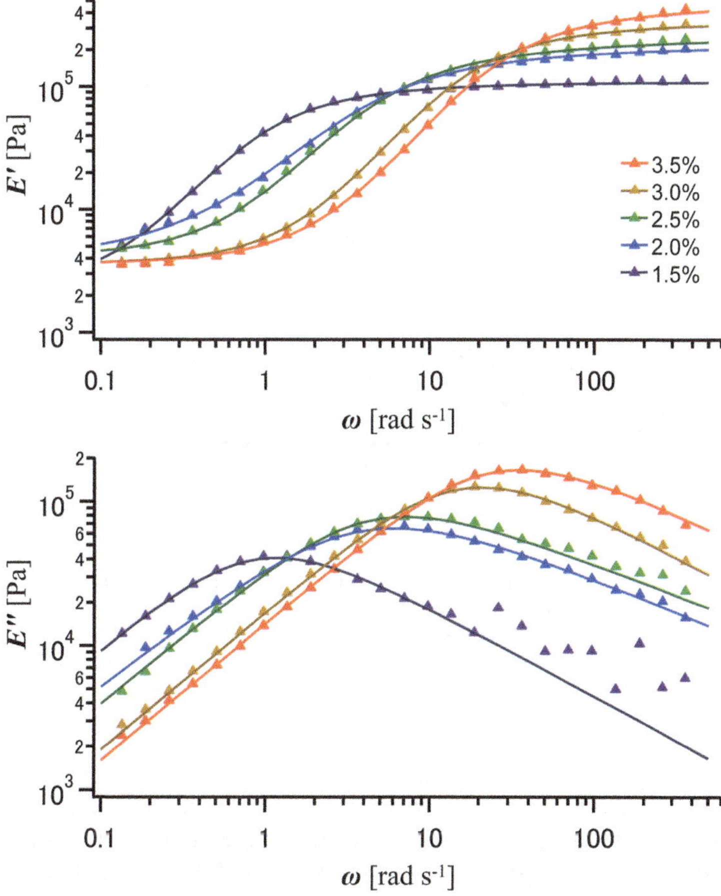

Figure 4.12   Frequency ω dependence of the storage (E') and loss (E'') moduli for polybutadiene-based slide-ring gels at 313 K with different cross-linker concentrations. The solvent is dimethyl sulfoxide (DMSO).
Reproduced from ref. 44 by permission of The Royal Society of Chemistry.

polyrotaxane. On the other hand, if the same estimation is done with the modulus at a lower frequency limit instead, M_x becomes more than 100 000, which is much higher than the M_n of the precursor polyrotaxane. This indicates the unlikely picture that several polyrotaxanes form a single partial network strand in slide-ring gels. Therefore, the higher plateau should be assigned as the rubbery one, which means that the lower one is not explainable by the rubber elasticity.

The viscoelastic data have another feature in the dynamics. The relaxation time becomes shorter with the increase of the cross-linker concentration. Because the modulus at the rubbery plateau also increases, the cross-linking density is thought to increase with the cross-linker concentrations. Thus, the relaxation dynamics should have correlation with the cross-linking density

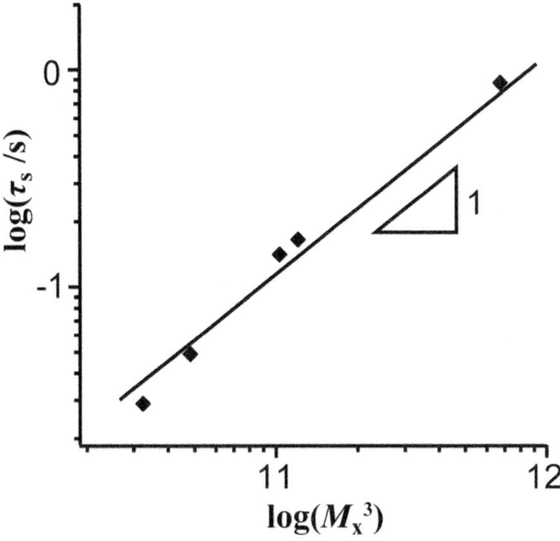

Figure 4.13 Cubic power dependence between the relaxation time (τ_s) and the averaged molecular weight of the network strands (M_x) for polybutadiene-based slide-ring gels.
Reproduced from ref. 44 by permission of The Royal Society of Chemistry.

and M_x. Indeed, the relaxation time τ_s bears a proportional relation to the cube of M_x (Figure 4.13). Similar cubic power dependence is widely observed in entangled polymer melts: the terminal (longest) relaxation time τ_d is proportional to the cube of the molecular weights of the whole chains M. The only difference in the case of the observed relaxation of the slide-ring gels is M_x, not M. Entangled polymers relax the anisotropic orientation of chain segments by reptation, and thus the dependence becomes stronger than unentangled polymers, which show weaker power dependence: $\tau_d \approx M^2$. This is because the entangled polymer chains cannot diffuse in any direction, but can diffuse only in the direction of the polymer axis, which is known as reptation. The cubic power dependence in slide-ring gels suggests that such strong constraint of diffusion is also imposed on the chain sliding through the cross-links. It can be regarded that the chain sliding is a kind of reptation. On the other hand, the cross-links in slide-ring gels are considerably sparser than the entangled points in polymer melts. Each partial chain, which is a chain between cross-links, in a slide-ring gel may be relaxed by the chain sliding individually to relax all the chains, whereas the entangled chains in polymer melts should diffuse with the distance of their own chain lengths. Therefore, polymer chains in the slide-ring gels can relax faster than entangled polymer melts, and the relaxation time depends on the cross-linking density. It is notable that the chain sliding can induce relaxation of the orientation anisotropy of the chains, but does not allow diffusion of the whole chain because of the end-capping groups. After the relaxation, another plateau appears with the equilibrium elasticity that originates from the entropy of the rings (see Section 4.3).

In this way, the characteristic relaxation observed in the slide-ring gels is induced by the chain sliding through the cross-links. This means that the relaxation time represents the time scale of the chain sliding. Thus, if a molecular design can achieve the control of the sliding dynamics, it can also influence the macroscopic relaxation time. One example is slide-ring gels that consist of different sized cyclodextrins.[46] Figure 4.14 shows the dependence of the relaxation time on normalized M_x for two series of slide-ring gels that were prepared from a polyrotaxane with α-cyclodextrin (α-polyrotaxane gels) or that with γ-cyclodextrin (γ-polyrotaxane gels). Both series of the slide-ring gels commonly exhibit similar viscoelastic relaxations with similar moduli at the rubbery plateau (100–300 kPa) and the equilibrium moduli (5–15 kPa). Similar to the above-mentioned polybutadiene-based slide-ring gels, these two gels also exhibit the cubic power dependence. This indicates that the observed relaxations are also attributed to the chain sliding through the cross-links, and that the relaxation times represent the sliding dynamics. The only difference is the relaxation time. The gels with larger cyclodextrins (γ-polyrotaxane gels) have considerably longer relaxation

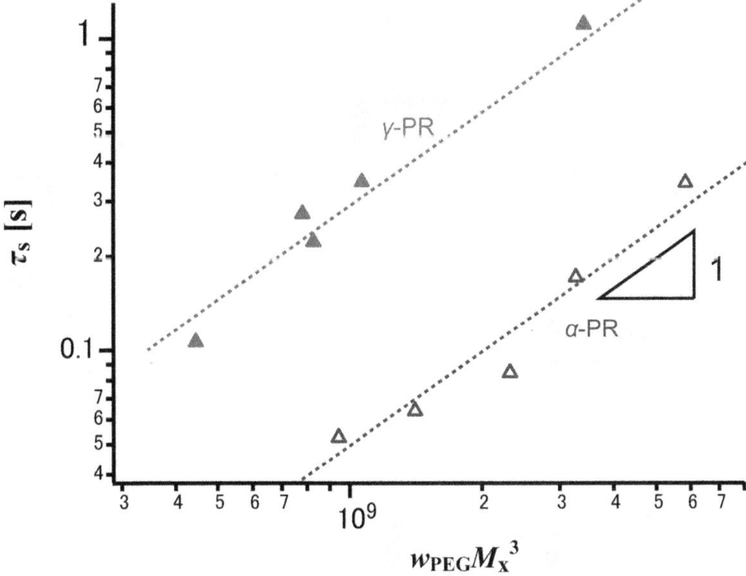

Figure 4.14 Cyclodextrin size effect on the relaxation dynamics of slide-ring gels that consist of PEG and α- or γ-cyclodextrin. The averaged molecular weight of network strands (M_x) is normalized with the weight fraction of the backbone (w_{PEG}).
Reproduced from ref. 46 by permission of The Royal Society of Chemistry.

times, suggesting that the chain sliding is much slower than those with smaller cyclodextrins (α-polyrotaxane gels). Although γ-cyclodextrin in the gel is threaded with a poly(ethylene glycol) chain, the large cavity has enough room to include other molecules such as the solvent. Molecular modeling indicates that the extra cavity is large enough to include two molecules of dimethyl sulfoxide (DMSO), the solvent of γ-polyrotaxane gels, and that the captured solvent can form CH–O interactions to bridge the inner surface of the cyclodextrin and poly(ethylene glycol). Thus, the interactions decelerate the chain sliding, and the corresponding macroscopic relaxation.

4.6   Biaxial Tensile Properties of Slide-ring Materials

Polymer gels exhibit unique mechanical properties such as low elastic moduli and large deformability owing to the entropic elasticity. The non-linear strain dependence of stress at moderate deformations has attracted the interest of many researchers in polymer science. In order to understand this complex dependence fully, it is necessary to characterize whole aspects of nonlinear elastic behavior. Uniaxial deformation (stretching or compression) has often been used because of its simplicity, but it is only a specific case among all physically accessible deformations. Analyses relying only on uniaxial data often lead to ambiguous conclusions. In contrast, general biaxial strain testing, in which the principal ratios in the two orthogonal directions are changed independently, can cover the whole range of admissible homogeneous deformations of incompressible materials. Therefore, it provides deterministic information on the strain energy density function of elastic materials. Urayama *et al.* have recently developed a novel biaxial tensile tester to detect quite low moduli (typically $\sim 10^2$ Pa) for extremely soft gels, including some biological soft tissues (Figure 4.15).[47] The state-of-the-art equipment enables them to investigate the nonlinear elasticity of the slide-ring gels, compared to usual chemical gels.

It has been empirically known that uniaxial stress–strain data for fully swollen gels and elastomers with chemical cross-links agree exceptionally well with predictions of the simplest model of rubber elasticity, *i.e.*, the ideal gas model or the neo-Hookean (NH) model. This feature is independent of the types of polymer, solvent, and cross-linking. The NH model assumes neither the interactions between different network strands nor the effects from the finite lengths of network strands. In fact, the nonlinear stress–strain curves of most real elastomers and gels deviate considerably from the NH model because of the presence of uncrossability due to trapped entanglements and finite extensibility. It is worth considering whether the exceptional agreement between uniaxial data and the NH model implies that fully swollen gels correspond to ideal gas systems, but the uniaxial testing only gives us limited information on the strain energy density function. Therefore, the biaxial data for fully swollen chemical gels with very low network densities are needed to understand the unsolved fundamental issue in the physics of rubber elasticity.

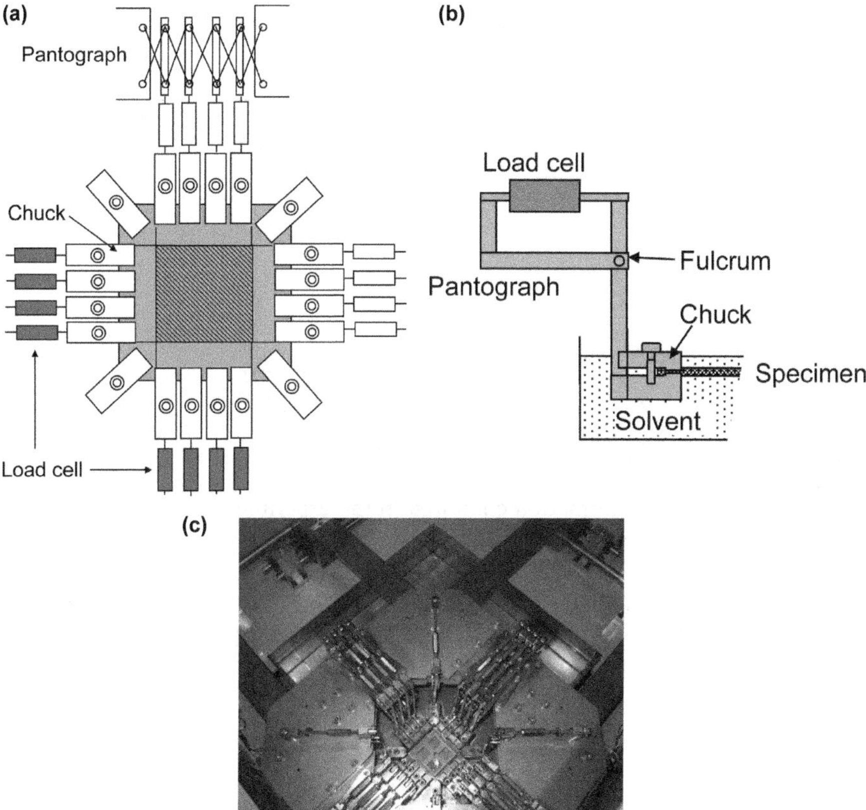

Figure 4.15 Schematic of the custom-built biaxial tensile tester: (a) top view, and (b) side view. The diagonal region corresponds to the effective specimen dimensions of the load. (c) Photograph of the custom-built biaxial tensile tester.
Reproduced from ref. 47 by permission of The Royal Society of Chemistry.

The three types of biaxial deformation (Figure 4.16), *i.e.*, equibiaxial stretching, pure shear deformation, and "two-step" biaxial deformation, are used to determine the strain energy density function. The equibiaxial stretching indicates that the sample sheets are equally stretched along the two orthogonal directions $(\lambda_x = \lambda_y)$, where λ_i $(i = x, y)$ is the ratio of the dimensions in the undeformed and deformed states along the i axis. In the pure shear deformation, the film specimen is uniaxially stretched in the x direction, while prohibiting lateral shrinking in the y direction $(\lambda_y = 1)$. The two-step biaxial deformation means that the gel is first deformed by pure shear up to $\lambda_x = \lambda_c$ and then stretched in the y direction, while maintaining $\lambda_x = \lambda_c$ until the equibiaxially stretched state of $\lambda_x = \lambda_y = \lambda_c$ is achieved. From these three types of biaxial deformation, we can obtain a whole profile of the strain energy density function in detail.

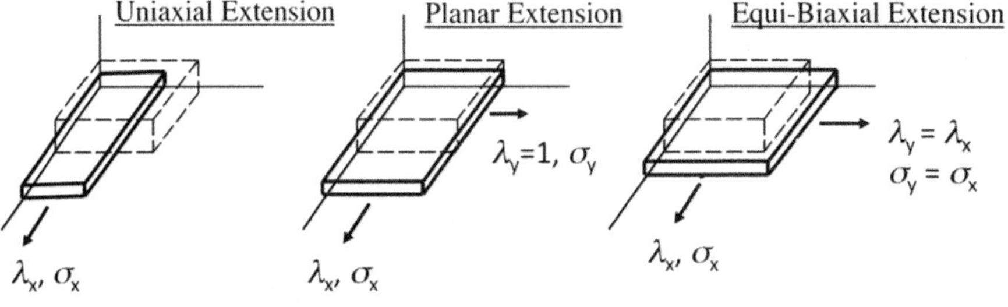

<u>Two-Step Extension</u>

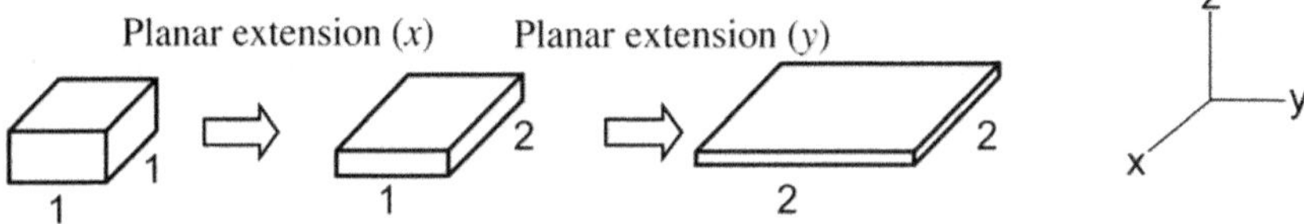

Figure 4.16   Types of deformation employed in the experiments.[48] Reproduced by permission of AIP Publishing.

The strain energy density function (F) describes the entire stress–strain behavior of elastic materials.[48] In general, F for incompressible elastomers ($\lambda_x\lambda_y\lambda_z = 1$) is expressed using the first and second strain invariants of deformation tensor I_i ($i = 1, 2$) as variables:

$$I_1 = \lambda_x^2 + \lambda_y^2 + \lambda_z^2 \tag{4.22a}$$

$$I_2 = \lambda_x^2\lambda_y^2 + \lambda_y^2\lambda_z^2 + \lambda_z^2\lambda_x^2. \tag{4.22b}$$

The stress σ_i in biaxial deformation is given by:

$$\sigma_i = \frac{2}{\lambda_i}\left(\lambda_i^2 - \frac{1}{\lambda_i^2\lambda_j^2}\right)(F_1 + \lambda_j^2 F_2) \quad (i,j = x,y), \tag{4.23}$$

where F_1 and F_2 are the derivatives of F with respect to I_1 and I_2, respectively, i.e., $F_1 \equiv \partial F/\partial I_1$ and $F_2 \equiv \partial F/\partial I_2$. Eqn (4.24) gives each derivative as a function of λ_i and σ_i ($i,j = x, y$).

$$F_1 = \frac{1}{2(\lambda_x^2 - \lambda_y^2)}\left[\frac{\lambda_x^3\sigma_x}{\lambda_x^2 - (\lambda_x\lambda_y)^{-2}} - \frac{\lambda_y^3\sigma_y}{\lambda_y^2 - (\lambda_x\lambda_y)^{-2}}\right] \tag{4.24a}$$

$$F_2 = \frac{-1}{2(\lambda_x^2 - \lambda_y^2)}\left[\frac{\lambda_x\sigma_x}{\lambda_x^2 - (\lambda_x\lambda_y)^{-2}} - \frac{\lambda_y\sigma_y}{\lambda_y^2 - (\lambda_x\lambda_y)^{-2}}\right]. \tag{4.24b}$$

Using eqn (4.24), even when the concrete form of F is unknown, F_1 and F_2 are obtained from the biaxial stress–strain data of $\lambda_x \neq \lambda_y$.

Urayama *et al.* compared slide-ring (SR) gels, Tetra-PEG gels, and usual chemical gels [poly(acrylamide), PAAm] in the stress ratio σ_x/σ_y in the first

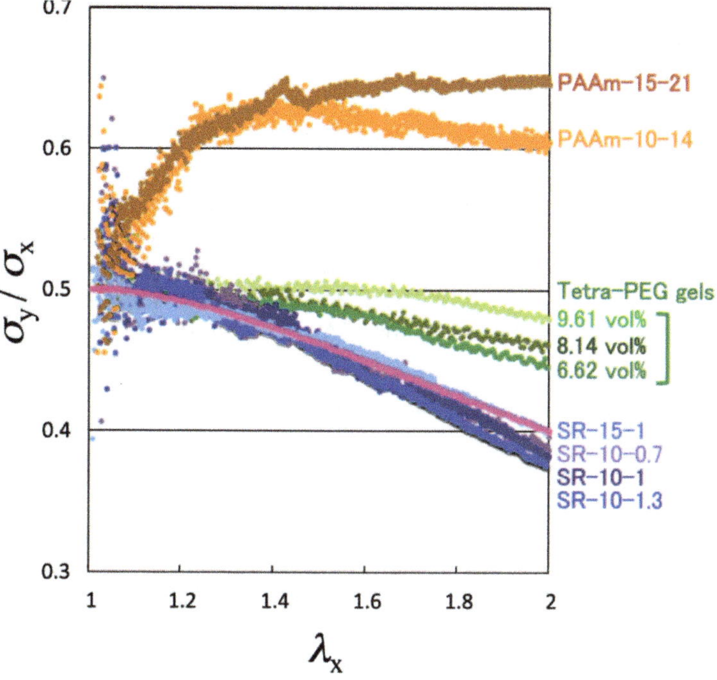

Figure 4.17 The ratio of the stresses in the stretched (x) and constrained (y) directions in planar extension as a function of the elongation ratio (λ_x) for the slide-ring (SR) gels, Tetra-PEG gels, and PAAm gels, where the red line represents the theoretical curve from eqn (4.25).[48] Reproduced by permission of AIP Publishing.

planar extension of the two-step deformation, as shown in Figure 4.17.[48] According to the infinitesimal elasticity theory, the stress ratio σ_y/σ_x in the zero-strain limit corresponds to the Poisson's ratio, which is around 0.5 in these gels. In the finite strain range of $\lambda_x > 1.2$, the data of all the SR gels almost collapse into a single curve independent of the concentrations of polyrotaxane and cross-linker. The stress ratio decreases with increasing λ_x, and becomes less than 0.4 at $\lambda_x = 2$. In contrast, σ_y/σ_x for the PAAm gels increases with stretching, and becomes almost constant around 0.65 in the region of $\lambda_x > 1.4$. The as-prepared Tetra-PEG gels with various network concentrations, which are chemical gels with nearly homogeneous structure, also indicate that the σ_y/σ_x ratios slightly decrease with stretching, and lie in the range between 0.45 and 0.5. Notably, the difference in the results between the SR gels and Tetra-PEG gels directly reflects the effect of slidability of cross-links because these gels are similar in the network backbone composed of poly(ethylene glycol). The ratio σ_y/σ_x in the planar extension suggests the strength of the influence of the strain in one direction on the stress in the other direction: the larger σ_y/σ_x (≤ 1) becomes, the stronger this cross-effect is. The results in Figure 4.17 indicate that this cross-effect is considerably smaller in the slide-ring gels than that in the chemical gels.

Figure 4.18 and Figure 4.19 show the I_1 dependence of the derivatives F_1 and F_2 for the slide-ring gels and PAAm-15-21, respectively, which was obtained from the results of the two-step deformation. The derivatives exhibit an apparent upswing or downswing in the vicinities of the two limiting states of $\lambda_x = \lambda_y$, *i.e.*, the undeformed state ($I_1 = 3$) and the equibiaxially stretched state (*i.e.*, $I_1 \approx 8$), where each derivative is undefined by eqn (4.24) due to the zero values in the denominator and numerator. These figures indicate that F_2 for all slide-ring gels is very small, and practically undetectable through the entire range of I_1, whereas F_2 for PAAm-15-21 is finite (Figure 4.19). This

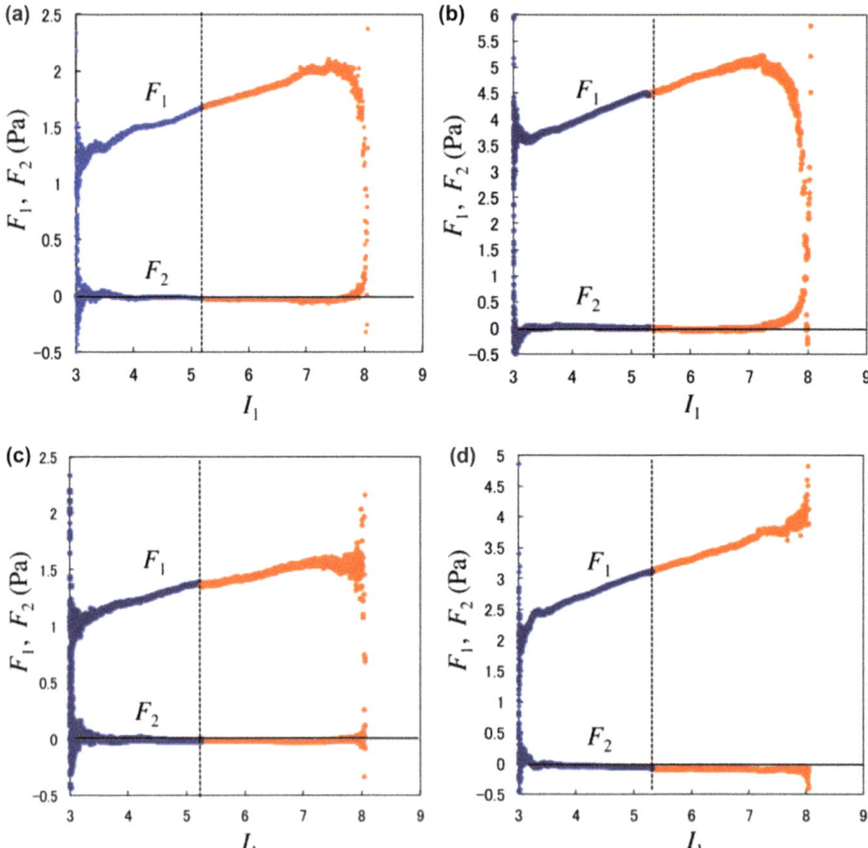

Figure 4.18   The derivatives of the strain free energies (F) with respect to I_1 and I_2 (F_1 and F_2, respectively) as a function of I_1 obtained from the stress–strain data of the two-step biaxial deformation for four kinds of slide-ring (SR) gels with different polyrotaxane concentrations on gelation and cross-linker concentrations such as SR-(polyrotaxane concentraition)-(cross-linker concentration): (a) SR-10-1, (b) SR-15-1, (c) SR-10-0.7, and (d) SR-10-1.3.[48]
Reproduced by permission of AIP Publishing.

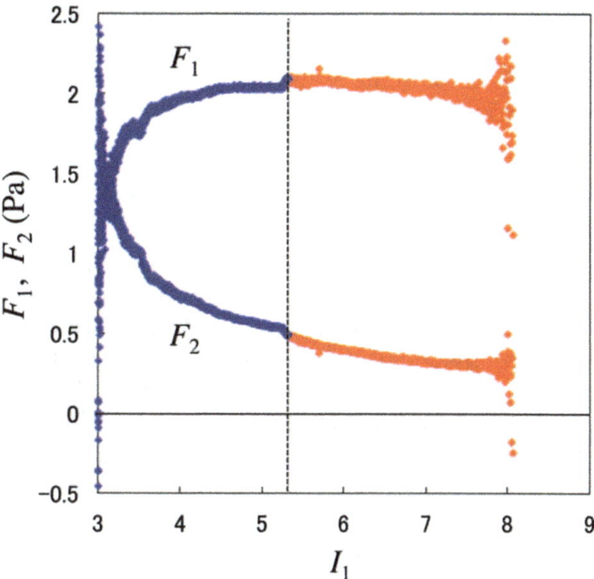

Figure 4.19   The derivatives of F with respect to I_1 and I_2 (F_1 and F_2, respectively) as a function of I_1, obtained from the stress–strain data of the two step biaxial deformation for PAAm-15-21.[48]
Reproduced by permission of AIP Publishing.

means that the strain energy density F of the slide-ring gels is a function of only I_1. This feature is unusual because F for most chemical gels and rubbers involves the terms of I_2 ($I_2 \neq 0$). Incidentally, Urayama *et al.* also revealed that F_2 for the chemical gels and elastomers does not vanish, even in the fully swollen state, which suggest that the biaxial strain properties do not obey the simple NH model.[48]

When $F_2 = 0$ in planar extension ($\lambda_y = 1$), eqn (4.23) gives the stress ratio σ_y/σ_x simply as:

$$\frac{\sigma_y}{\sigma_x} = \frac{\lambda_x}{\lambda_x^2 + 1}. \tag{4.25}$$

As shown in Figure 4.17, the data of all the slide-ring gels almost obey eqn (4.25), depicted by the solid line. This means that the features of $F_2 = 0$ and the small stress ratio in planar extension are equivalent. Since I_2 represents the cross-effect of the strains in different directions, as is evident from the definition [eqn (4.22b)], the absence of the I_2 term in the strain energy density function $F(F_2 = 0)$ directly indicates that there is no explicit cross-effect of the strains in different directions. This characteristic explains the feature of the small stress ratio in planar extension, *i.e.*, the small effect of the strain in one direction and on the stress in the other direction. The upper region beyond the solid line [eqn (4.25)] in Figure 4.17 corresponds to the condition of $F_2 > 0$. This straightforwardly suggests that a positive cross effect of the strains in different directions enhances the stress ratio σ_y/σ_x. Most of the chemical gels and elastomers belong to this category.

The absence of the I_2 term in the strain energy density function F suggests that the biaxial strain behaviors of slide-ring gels can obey the NH model if the effect of finite extensibility is taken into account. The Gent model is a function of only I_1, including the finite extensibility effect as:

$$F(I_1) = -\frac{G}{2}(I_m - 3)\ln\left(1 - \frac{I_1 - 3}{I_m - 3}\right), \tag{4.26}$$

where G is the initial shear modulus, and I_m is the value of I_1 at ultimate deformation, at which the network strands are fully stretched. The stress diverges at $I_1 = I_m$, and the NH model corresponds to the case of $I_m = \infty$. The classical three chain model using an inverse Langevin function is well known as a simple extension of the NH model to include the finite extensibility effect. However, F_2 of the three chain model is finite, which does not satisfy the hypothesis of $F_2 = 0$. The Gent model gives us the biaxial stress–strain relations as:

$$\sigma_i = G\left(\lambda_i - \frac{1}{\lambda_i^3 \lambda_j^2}\right)\frac{I_m - 3}{I_m - I_1} \quad (i,j = x,y). \tag{4.27}$$

The third factor $(I_m - 3)/(I_m - I_1)$ represents the finite extensibility effect. The Gent model satisfactorily describes the stress–strain data in each deformation for all slide-ring gels.

It should be noted again that the strain energy density function F for most chemical gels and elastomers explicitly involves the I_2 terms, and that the stress–strain behavior is not described by the Gent model. For instance, biaxial stretching experiments revealed that the strain energy density function F for the Tetra-PEG gels, one of the nearly homogeneous chemical gels, is expressed by the sum of the Gent model and a linear term of I_2. The absence of an I_2 term in the strain energy density function F is a unique feature of the slide-ring gels, which originates from the movable cross-links along the network strands. The network structure in the deformed slide-ring gels is varied by the movable cross-links so that the entropy of the network configuration, including the one-dimensional distribution of cross-linked and uncross-linked cyclodextrins along the network strands, can be maximized. This function considerably homogenizes the network configuration in the deformed state, which leads to a significant decoupling between the imposed strains in two different directions. The implicit and minimal cross-effect of the strains in the slide-ring gels arises only from the volume constancy during deformation ($\lambda_x \lambda_y \lambda_z = 1$).

The Gent model describes the features of the finite deformation behavior of the slide-ring gels phenomenologically, although the effect of movable cross-links on nonlinear elasticity is not explicitly modeled with molecular basis. It remains a challenging problem to build a molecular statistical model explaining the nonlinear elasticity of the slide-ring gels. The biaxial tensile properties of the as-prepared slide-ring gels will provide definite experimental data compared to the molecular statistical models.

4.7 Stretch-driven Volume Change of the Slide-ring Gels

The stretch-driven volume change is a general property characteristic of gels in a semi-open system, where the solvent can flow in and out of the gels. When a one-dimensional constant tensile strain is imposed externally to gels that are fully swollen in solvents, they further swell by a force to increase the conformational entropy (*i.e.*, reduce the anisotropy in conformation) of the deformed networks by the tensile stress, as shown in Figure 4.20. The volume change before and after imposing a constant strain in the fully swollen state has been explainable by the Flory–Rehner model,[14] where the swelling equilibrium was achieved by a balance between the osmotic swelling pressure (π_{mix}) originating from the isotropic mixing of network and solvent, and the elastic resistive stress (π_{el}) from the entropic elasticity of the network strands. Urayama *et al.* extended the model to a new one based on the scaling approach for the elastic term of the free energy of a network (F_{el}) and the mixing (osmotic) term (F_{mix}) to consider the concentration fluctuation in the semi-dilute regime.[49] As a result, a simple relation between the stretch ratios $\alpha_\parallel$ and $\alpha_\perp$ in the directions longitudinal and transverse to the tensile strain, respectively, was obtained

$$\alpha_\perp = \alpha_\parallel^{-(1-\nu_e)/(1+\nu_e)}, \tag{4.28}$$

where ν_e is the excluded volume exponent depending on the solvent, and $\alpha_\parallel$ represents the stretch ratio induced by the tensile strain. Since the

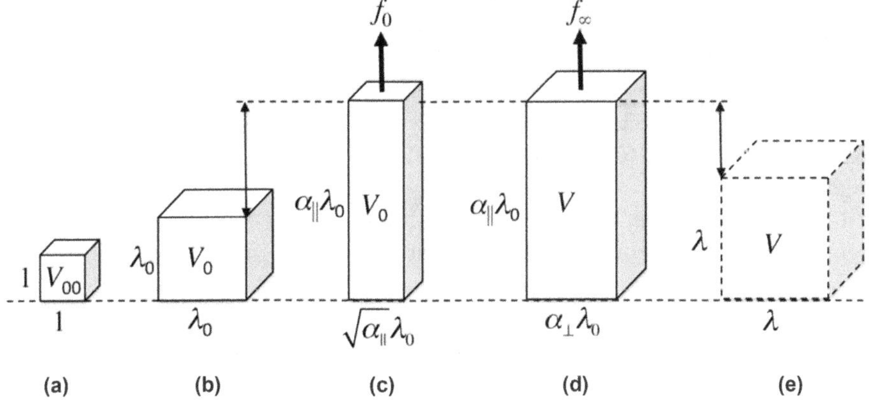

Figure 4.20 Schematic diagram of the stretch-driven swelling of gels. When a one-dimensional constant tensile strain is imposed externally to gels that are fully swollen in solvents, they further swell by a force to increase the conformational entropy of the deformed networks by the tensile stress: (a) reference state, (b) equilibrium swollen state with a volume of V_0, (c) externally imposed stretching $\alpha_\parallel$, (d) externally imposed stretching $\alpha_\perp$ in the equilibrium swollen state, where the stretching induces further swelling V ($> V_0$), and (e) isotropic swollen state with a volume of V.[50] Reprinted with permission from *Macromolecules*. Copyright (2012) American Chemical Society.

equilibrium Poisson's ratio μ_∞ is defined by the ratio of the true strains (ε) in the longitudinal and transverse directions:

$$\mu_\infty = -\frac{\varepsilon_\perp}{\varepsilon_\parallel} = -\frac{\ln \alpha_\perp}{\ln \alpha_\parallel}. \tag{4.29}$$

Eqn (4.29), a generalized definition of Poisson's ratio, is valid in finite deformation, as well as in the small deformation satisfying the linear elasticity. Consequently, μ_∞ is given by:

$$\mu_\infty = \frac{1 - \nu_e}{1 + \nu_e}. \tag{4.30}$$

The equilibrium Poisson's ratio μ_∞ indicates a measure of the degree of strain-induced swelling, because the equilibrium volumes of gels before and under stretching (designated as V_0 and V, respectively) are given by:

$$\frac{V}{V_0} = \frac{\phi_0}{\phi} = \alpha_\parallel^{1 - 2\mu_\infty}, \tag{4.31}$$

where ϕ_0 and ϕ are the volume fractions of the gels before and under stretching, respectively.

As is evident from eqn (4.30), μ_∞ depends only on ν_e, *i.e.*, the solubility of the constituent polymers in solvents: $\mu_\infty = 1/4$ for typical good solvents with $\nu_e = 3/5$, and $\mu_\infty = 1/3$ for theta solvents with $\nu_e = 1/2$. It is to be noted that μ_∞ is independent of the degree of deformation $\alpha_\parallel$, as well as the structural parameters of gels (such as the length of the network strands, and preparation concentration). Interestingly, the value of $\mu_\infty = 1/4$ for good solvents is identical to the result obtained by the classical theory (Flory–Rehner model) where the excluded volume effect is not considered. This accordance comes from the cancellations of the absences of excluded volume effect in both F_{el} and F_{mix} in the classical theory. The $\alpha_\parallel$ independence of μ_∞ is in a good agreement with the experimental results of chemical gels, as shown in Figure 4.21.

In contrast, Figure 4.22 shows that slide-ring gels have a definite strain dependence of the equilibrium (osmotic) Poisson's ratio (μ_∞). The stretch ($\alpha_\parallel$) dependence of μ_∞ for the slide-ring gels indicates a crossover at $\alpha_\parallel = \alpha_c$; μ_∞ increases with $\alpha_\parallel$ in moderate stretching of $\alpha_\parallel < \alpha_c$, whereas μ_∞ levels off at high stretching of $\alpha_\parallel < \alpha_c$. The $\alpha_\parallel$-dependent μ_∞ under moderate stretching is attributable to the structural homogenization by the pulley effect, which suppresses the subsequent stretch-induced swelling. On the other hand, $\alpha_\parallel$-independent μ_∞ under high stretching may be explained by the suppression of the pulley effect because of a considerable localization and/or stacking of many free rings to the chain ends [Figure 4.23(c)].

The slide-ring gels have several structural parameters such as the cross-linker concentration, the filling ratio of cyclodextrins in polyrotaxanes, and the backbone chain length of the polyrotaxanes. These parameters may significantly affect the mobility of the cross-linking points, which leads to

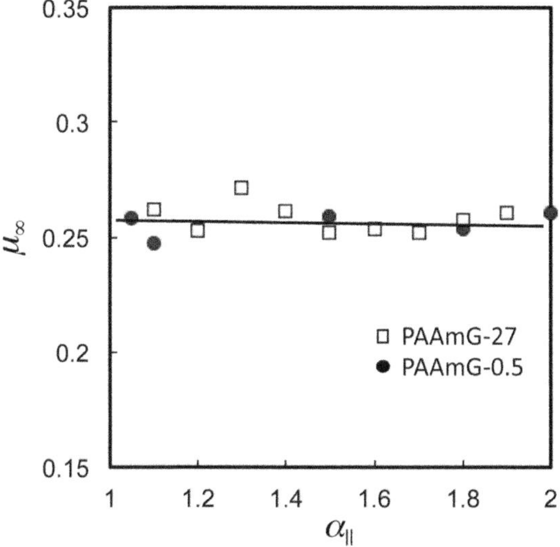

Figure 4.21　Equilibrium Poisson's ratio (μ_∞) as a function of imposed stretch $(\alpha_\parallel)$ for polyacrylamide hydrogels. The value $(\mu_\infty \approx 0.255)$ is independent of $\alpha_\parallel$, as well as cross-link concentration.[50]
Reprinted with permission from *Macromolecules*. Copyright (2012) American Chemical Society.

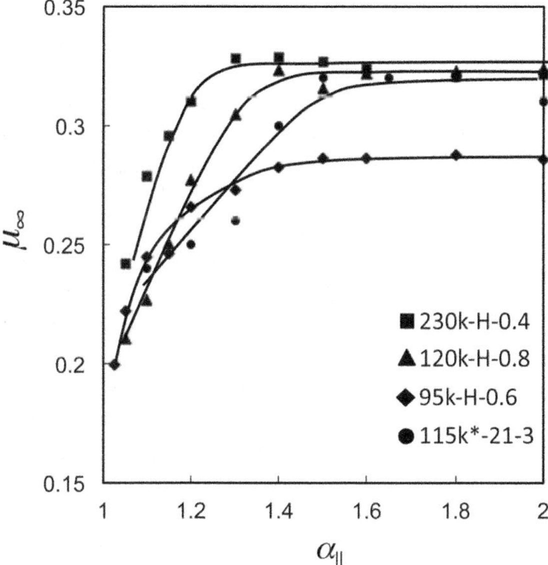

Figure 4.22　Equilibrium Poisson's ratio (μ_∞) as a function of imposed stretch $(\alpha_\parallel)$ for the slide-ring gels with various lengths of polyrotaxane strands showing the type A behavior.[50]
Reprinted with permission from *Macromolecules*. Copyright (2012) American Chemical Society.

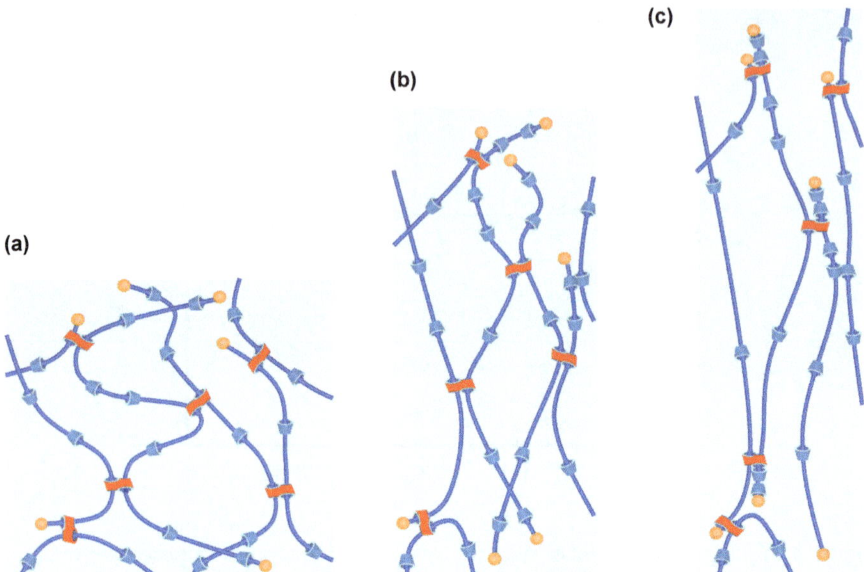

Figure 4.23   Schematics of polyrotaxane gels with slidable cross-links (a) in the relaxed state, (b) under moderate stretching, and (c) under high stretching.[49]
Reprinted with permission from *Macromolecules*. Copyright (2012) American Chemical Society.

the stretch-driven volume change of the slide-ring gels. Urayama *et al.* systematically investigated the dependence of the stretch-driven swelling properties on cross-linker concentration, backbone chain length, and the inclusion ratio.[50] For high cross-linker concentrations, μ_∞ is independent of $\alpha_\parallel$, which is similar to the behavior of the usual chemical gels (PAAm gels) with fixed cross-links (type B). When cross-linker concentrations are sufficiently low, μ_∞ becomes dependent on $\alpha_\parallel$: at the moderate stretching of $\alpha_\parallel < \alpha_c$, μ_∞ steeply increases with $\alpha_\parallel$, whereas at the high stretching of $\alpha_\parallel < \alpha_c$, μ_∞ reaches a plateau value (type A), or μ_∞ gradually increases with $\alpha_\parallel$ (type A*), as shown in Figure 4.24. Since the model assuming the affine displacement of cross-links predicts the $\alpha_\parallel$-independent μ_∞, the $\alpha_\parallel$-dependent μ_∞ of the slide-ring gels under moderate stretching is attributed to the nonaffine displacement of cross-linked cyclodextrins *via* the pulley effect. The stretch-driven swelling arises from an entropic force to decrease the conformational anisotropy of the network strands due to imposed stretching. The slide-ring gels with highly mobile rings originally have the pulley effect to homogenize the network topology under imposed strain. Ideally, the pulley effect passes through the network strands to homogenize the tension on them in response to applied deformation so that the conformational entropy can be maximized. Consequently, the pulley effect should suppress the subsequent stretch-induced swelling, which leads to an increase of μ_∞ with increasing $\alpha_\parallel$. If the cross-linking junctions are immobile, μ_∞ would remain the value in the small-strain limit independently of $\alpha_\parallel$.

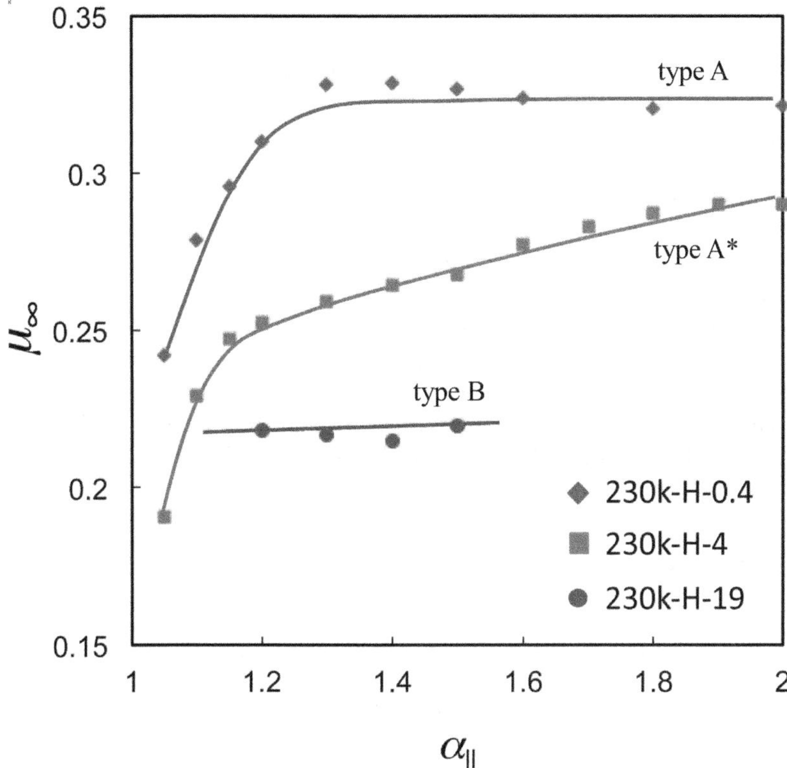

Figure 4.24 Equilibrium Poisson's ratio (μ_∞) as a function of imposed stretch ($\alpha_\parallel$) for slide-ring gels with various moduli. The lines are guides for eyes. The samples 230k-H-0.4, 230k-H-4, and 230k-H-19 show the type A, type A*, and type B behaviors, respectively.[50]
Reprinted with permission from *Macromolecules*. Copyright (2012) American Chemical Society.

At high elongation, many rings, including the uncross-linked free ones, would be localized and/or stacked to the chain ends, and lose mobility. As a result, the pulley effect of the slide-ring gels is considerably suppressed, and thereby μ_∞ becomes independent of $\alpha_\parallel$, as in the case of usual chemical gels with fixed cross-links. This scenario explains type A behavior.

If the number of cross-linking junctions or free rings is too large, the mobility of the rings is markedly suppressed due to the restriction of the slidable range along the strands. In such highly cross-linked slide-ring gels, the pulley effect does not work much less actively, which leads to the type B behavior.

The transition between type A and B behaviors by varying the cross-linker concentration is observed independently of M_{PEG} (molecular weight of backbone PEG), as well as the filling ratio ϕ_{CD}. The type A* behavior is seen for the gels made of the polyrotaxane with relatively high ϕ_{CD}, but it disappears for the gels made of the polyrotaxane with low ϕ_{CD} and almost the same M_{PEG}, which indicates that an important factor for the type A* behavior is ϕ_{CD} rather than M_{PEG}. The type A* behavior is detected at the moderate cross-linker concentrations between the regimes for types A and B. When ϕ_{CD} is low, the transition between types A and B directly occurs in a narrow Young's modulus (E) range of 6 kPa $< E <$ 13 kPa without type A*. The high ϕ_{CD} seems to broaden the transition between types A and B, and hence type A* can be regarded as an intermediate behavior between the two types. If the free cyclodextrins along the network strands have broad number distribution, the localization and/or stacking of cyclodextrins at the chain ends would proceed gradually in a finite range of stretch, which results in the type A* behavior.

The modeling of the pulley effect on μ_∞ remains a challenging issue. In particular, one should propose an alternative form of the free energy of a network (F_{el}) due to the nonaffine displacement of movable cross-links. It is necessary to characterize experimentally in detail the nonlinear elasticity of networks with movable cross-links, and to deduce the form of F_{el} based on the experimental data.

References

1. L. R. G. Treloar, in *The Physics of Rubber Elasticity*, Oxford University Press, Oxford, 3rd edn, 1975.
2. J. E. Mark and B. Erman, in *Rubber Elasticity: A Molecular Primer*, Cambridge University Press, Cambridge, 2nd edn, 2007.
3. W. Kuhn, *Colloid Z.*, 1936, **76**, 258.
4. H. M. James and E. Guth, *J. Chem. Phys.*, 1943, **11**, 455.
5. P. J. Flory and J. Rehner, Jr., *J. Chem. Phys.*, 1943, **11**, 512; P. J. Flory and J. Rehner, Jr., *J. Chem. Phys.*, 1943, **11**, 521.
6. S. F. Edwards and K. F. Freed, *J. Phys.*, 1970, **C3**, 739; S. F. Edwards and K. F. Freed, *J. Phys.*, 1970, **C3**, 750; S. F. Edwards and K. F. Freed, *J. Phys.*, 1970, **C3**, 760.
7. H. Benoit, D. Decker, R. Duplessix, C. Picot, J. P. Cotton, B. Farnoux, G. Jarnick and R. Ober, *J. Polym. Sci., Polym. Phys. Ed.*, 1976, **14**, 2119.
8. J. A. Hinkley, C. C. Han, B. Mozer and H. Yu, *Macromolecules*, 1978, **11**, 836.
9. T. Sakai, T. Matsunaga, Y. Yamamoto, C. Ito, R. Yoshida, S. Suzuki, N. Sasaki, M. Shibayama and U. I. Chung, *Macromolecules*, 2008, **41**, 5379.
10. Y. Akagi, T. Katashima, Y. Katsumoto, K. Fujii, T. Matsunaga, U. Chung, M. Shibayama and T. Sakai, *Macromolecules*, 2011, **44**, 5817.
11. P. G. de Gennes, in *Scaling Concept in Polymer Physics*, Cornell University Press, Ithaca, 1979.
12. M. Doi and S. F. Edwards, in *The Theory of Polymer Dynamics*, Oxford University Press, Oxford, 1988.
13. R. C. Ball, M. Doi, S. F. Edwards and M. Warner, *Polymer*, 1981, **22**, 1010.
14. P. J. Flory and J. J. Rehner, *J. Chem. Phys.*, 1943, **11**, 521.
15. T. Tanaka, *Phys. Rev. Lett.*, 1978, **40**, 820.
16. T. Tanaka, D. J. Fillmore, S.-T. Sun, I. Nishio, G. Swislow and A. Shah, *Phys. Rev. Lett.*, 1980, **45**, 1636.

17. T. Tanaka and D. J. Fillmore, *J. Chem. Phys.*, 1979, **70**, 1214.
18. M. Shibayama, *Macromol. Chem. Phys.*, 1998, **199**, 1.
19. J. P. Gong, *Soft Matter*, 2006, **2**, 544.
20. J. P. Gong, Y. Katsuyama, T. Kurokawa and Y. Osada, *Adv. Mater.*, 2003, **15**, 1155.
21. J. P. Gong, *Soft Matter*, 2010, **6**, 2583.
22. R. Yoshida, *Biophysics*, 2012, **8**, 163.
23. R. Yoshida, K. Uchida, Y. Kaneko, K. Sakai, A. Kikuchi, Y. Sakurai and T. Okano, *Nature*, 1995, **374**, 240.
24. *Gels Handbook*, ed. Y. Osada and K. Kajiwara, Academic Press, Elsevier, Amsterdam, 2000.
25. K. Haraguchi and T. Takeshita, *Adv. Mater.*, 2002, **14**, 1120.
26. J. Li, A. Harada and M. Kamachi, *Polym. J.*, 1994, **26**, 1019.
27. J. Watanabe, T. Ooya and N. Yui, *J. Artif. Organs.*, 2000, **3**, 136.
28. T. Takata, *Polym. J.*, 2006, **38**, 1.
29. Y. Okumura and K. Ito, *Adv. Mater.*, 2001, **13**, 485.
30. P. G. de Gennes, *Physica A*, 1999, **271**, 231.
31. S. Granick and M. Rubinstein, *Nat. Mater.*, 2004, **3**, 586.
32. K. Ito, *Polym. J.*, 2007, **39**, 488.
33. K. Ito, *Curr. Opin. Solid State Mater. Sci.*, 2010, **14**, 28.
34. R. S. Stein, *J. Chem. Educ.*, 1958, **35**, 203.
35. DoITPoMS, Elasticity in biological Materials, University of Cambridge, available at http//www.doitpoms.ac.uk/tlplib/bioelasticity/index.php (accessed June 2015).
36. J. F. V. Vincent, in *Structural Biomaterials*, Princeton University Press, Princeton, 1990.
37. S. Vogel, in *Comparative Biomechanics: Life's Physical World*, Princeton University Press, Princeton, 2003.
38. K. Mayumi and K. Ito, *Polymer*, 2010, **51**, 959.
39. K. Ito, *Polym. J.*, 2011, **44**, 38.
40. K. Mayumi, M. Nagao, H. Endo, N. Osaka, M. Shibayama and K. Ito, *Phys. B*, 2009, **404**, 2600.
41. M. B. Pinson, E. M. Sevick and D. R. M. Williams, *Macromolecules*, 2013, **46**, 4191.
42. A. Bando, K. Mayumi, H. Yokoyama and K. Ito, *React. Funct. Polym.*, 2013, **73**, 904.
43. K. Mayumi, M. Tezuka, A. Bando and K. Ito, *Soft Matter*, 2012, **8**, 8179.
44. K. Kato and K. Ito, *Soft Matter*, 2011, 7, 8737.
45. K. Kato, T. Yasuda and K. Ito, *Macromolecules*, 2013, **46**, 310.
46. K. Kato, K. Karube, N. Nakamura and K. Ito, *Polym. Chem.*, 2015, **6**, 2241.
47. B. Yohsuke, K. Urayama, T. Takigawa and K. Ito, *Soft Matter*, 2011, 7, 2632.
48. Y. Kondo, K. Urayama, M. Kidowaki, K. Mayumi, T. Takigawa and K. Ito, *J. Chem. Phys.*, 2014, **141**, 134906.
49. N. Murata, A. Konda, K. Urayama, T. Takigawa, M. Kidowaki and K. Ito, *Macromolecules*, 2009, **42**, 8485.
50. A. Konda, K. Mayumi, K. Urayama, T. Takigawa and K. Ito, *Macromolecules*, 2012, **45**, 6733.

Solid-state Properties of Polyrotaxanes

5.1 Structure and Viscoelastic Properties of Chemically Modified Polyrotaxanes in the Solid State

Polyrotaxane, which consists of cyclic molecules threaded onto a linear polymer chain in a mechanical interlocking manner, has been studied widely due to potential applications as smart and stimuli-responsive materials. Cyclic molecules can slide and rotate with spatial constraints on a backbone polymer string. The unique supramolecular structure of polyrotaxane enables us to apply it to various nanoscale molecular devices. One of the most used polyrotaxane in research consists of α-cyclodextrin as a ring molecule and poly(ethylene glycol) as a backbone string because it is rather easy to synthesize the interlocking structure. A significant feature of the polyrotaxane is that 18 hydroxyl groups of α-cyclodextrins can be modified with different functional groups such as hydroxypropyl, methyl, and acetyl moieties. Such chemically modified polyrotaxanes (*i.e.*, polyrotaxane derivatives) exhibit improved characteristics, such as solubility in organic solvents, as well as unique functions of thermoresponsivity,[1,2] photoresponsivity,[3] and so on. In addition, polyrotaxane derivatives have also promoted research of the solid-state properties of polyrotaxanes, and the applications of polyrotaxanes to elastomers, resins, and composites in a solvent-free system. Kidowaki *et al.* reported that a liquid crystalline polyrotaxane (LCPR) having mesogenic side chains on the cyclodextrins behaved as a thermotropic liquid crystal.[4] And Araki *et al.* synthesized a "sliding graft copolymer (SGC)," which has side chains attached to cyclodextrins that are freely mobile along and rotatable around the main chain.[5] A cross-linked SGC film prepared by connecting the ends of the side chains, a caprolactone

Monographs in Supramolecular Chemistry No. 15
Polyrotaxane and Slide-Ring Materials
By Kohzo Ito, Kazuaki Kato and Koichi Mayumi
© Kohzo Ito, Kazuaki Kato and Koichi Mayumi 2016
Published by the Royal Society of Chemistry, www.rsc.org

origomer, to each other showed scratch-resistant properties, which can be applied to self-healing coatings, as mentioned in Section 8.4. In addition, polyrotaxane–cellulose hybrid fibers achieved high-tensile properties.[6]

The structure and dynamics of polyrotaxane derivatives in the solid state are quite important from the viewpoint of the applications of novel polymeric materials. Inomata *et al.* investigated the physical properties of a series of hydroxypropylated polyrotaxanes as polyrotaxane derivatives in the solid state systematically, including the crystallinity of the packing structure for cyclodextrins in unmodified polyrotaxane and polyrotaxane derivatives in the solid state, and the slow dynamics related to mechanical relaxation processes in solid-state polyrotaxanes.[7] The hydroxypropylation of polyrotaxane hinders the hydrogen bonds among the cyclodextrins, and thereby prevents aggregation of the cyclodextrins in polyrotaxane and among different polyrotaxanes, which substantially improves the solubility of polyrotaxane in a variety of solutions, especially aqueous ones.[6] The hydroxypropylation of polyrotaxane also reduces the interaction between cyclodextrins, and in the solid state hinders the packing structure of cyclodextrins similarly.

Figure 5.1 shows the wide-angle X-ray scattering (WAXS) profiles of unmodified polyrotaxane, hydroxypropylated polyrotaxanes, and poly(ethylene glycol) at room temperature.[7] All profiles are largely different from those of the cage-type crystalline structures of α-cyclodextrin without guest molecules in the cavities. Unmodified polyrotaxane demonstrates strong diffraction peaks (Figure 5.1(a)) at $q = 0.908$, 1.39, and 1.58 Å^{-1}, where q is the magnitude of the scattering vector given by $q = 4\pi\sin\theta/\lambda$ (2θ is the scattering angle, and λ is the wavelength of an X-ray, 1.5418 Å). These peaks were assigned to 110, 210, and 300 reflections, respectively, for the hexagonal lattice of cyclodextrins in a channel-type crystalline structure (Figure 5.2(a)) due to the hydrogen bonding between cyclodextrins. Figure 5.1 shows that reflections only from the regularity in the *ab* plane of the hexagonal packing are observed. This suggests that cyclodextrins in the polyrotaxane are placed in hexagonally packed columns with a disordered arrangement in the *c* axis, as schematically in Figure 5.2(b). The poorly ordered arrangement of cyclodextrins along the *c* axis may arise from a low filling ratio around 25% of cyclodextrins in the polyrotaxane.

In contrast, all WAXS profiles of the hydroxypropylated polyrotaxanes (Figure 5.1(b)–(e)) commonly demonstrate amorphous halos. This means that chemically modified cyclodextrins form no crystalline structures with highly ordered packing owing to suppression of intermolecular hydrogen bonding between the cyclodextrin hydroxyl groups, as well as steric hindrance by the hydroxypropyl groups. Small peaks at $q = 1.35$ and 1.64 Å^{-1} overlapping the amorphous halo in the WAXS profiles correspond to the diffraction peaks for a crystalline structure of pure poly(ethylene glycol) (Figure 5.1(f)). This means that some parts of uncovered poly(ethylene glycol) segments in hydroxypropylated polyrotaxanes seem to form a crystalline structure similar to pure poly(ethylene glycol). Such crystalline peaks are not

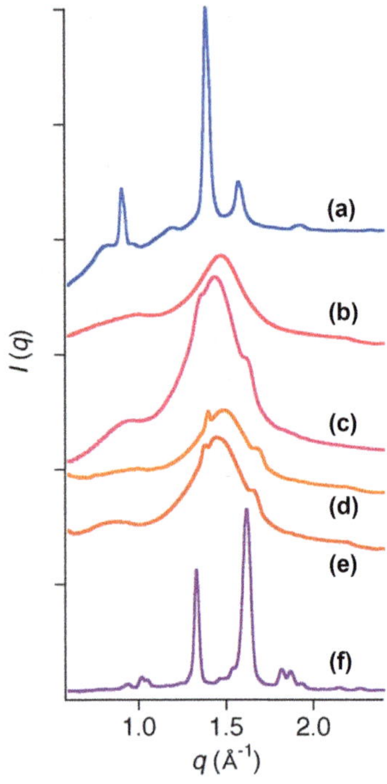

Figure 5.1   Wide-angle X-ray scattering (WAXS) profiles of (a) polyrotaxane (PR),
(b) hydroxypropylated polyrotaxane (HyPR) HyPR25-28, (c) HyPR38-28,
(d) HyPR40-21, (e) HyPR78-21, and (f) poly(ethylene glycol) (PEG) PEG35.
The HyPR sample names represent the modification and filling ratios.
For example, HyPR25-28 was named after a modification ratio of 25%
and filling ratio of 28%.[7]
Reprinted with permission from *Macromolecules*. Copyright (2010)
American Chemical Society.

observed for the hydroxypropylated polyrotaxane with a lower modification
ratio (Figure 5.1(b)), which suggests that interactions between cyclodextrins,
such as hydrogen bonds between the unmodified hydroxyl groups in
the hydroxypropylated polyrotaxane, may prevent the crystallization of
poly(ethylene glycol) segments effectively.

Figure 5.3 summarizes the differential scanning calorimetry (DSC) measurement results for unmodified polyrotaxane, and all of the hydroxypropylated polyrotaxanes. Unmodified polyrotaxane has no thermal
transitions over the measured temperature range and has thermal decomposition at higher temperature. This means that neat polyrotaxane
keeps the channel crystalline structure of cyclodextrins at temperatures
below the oxidative degradation temperature of poly(ethylene glycol). And
the thermal transitions of melting and crystallization of poly(ethylene glycol)
also disappear in the DSC profiles. On the other hand, the DSC profiles for
the cooling runs of some hydroxypropylated polyrotaxanes demonstrated
slight baseline shifts, as shown by the dashed arrows. This indicates the

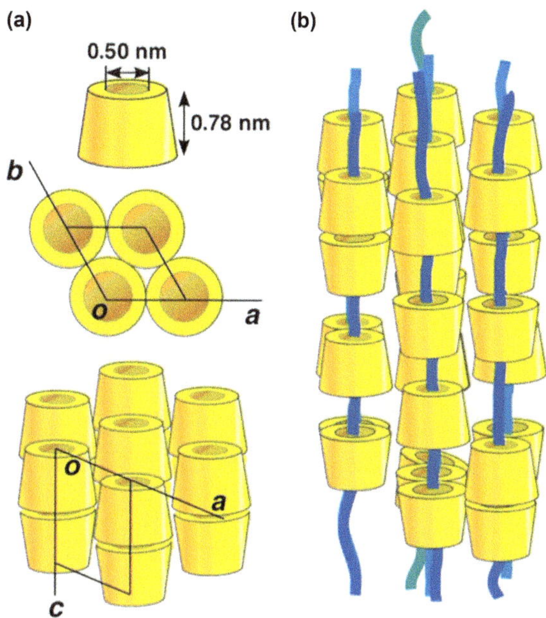

Figure 5.2 (a) Channel crystalline structure of α-cyclodextrins. (b) Hexagonally packed columns formed by α-cyclodextrins in polyrotaxane with disordered arrangement in the *c* axis.[7]
Reprinted with permission from *Macromolecules*. Copyright (2010) American Chemical Society.

glass transition of general amorphous polymers, of which temperatures are much higher than the glass transition temperature of pure poly(ethylene glycol) (*ca.* $-51\,^\circ$C). And all of the hydroxypropylated polyrotaxane samples showed endothermic peaks at 50–60 $^\circ$C on the heating runs, which are close to the melting point of pure poly(ethylene glycol). Hence, the transition probably results from the melting of the poly(ethylene glycol) crystalline domain in hydroxypropylated polyrotaxanes. These interpretations are supported by the temperature dependence of the WAXS profiles.

Inomata *et al.* also measured the viscoelastic profiles of unmodified polyrotaxane and polyrotaxane derivatives.[7] Unmodified polyrotaxane shows no frequency dispersion, which is consistent with the lack of thermal phase transitions below the decomposition temperature (Figure 5.4). In contrast, some hydroxypropylated polyrotaxanes clearly exhibit viscoelastic relaxation processes. To analyze viscoelastic profiles at different temperatures, the time–temperature superposition principle is widely used in polymer science. The data at different temperatures can be superimposed by shifting the angular frequencies, and the shift factor gives us the activation energy of the relaxation process. Figure 5.4(c) and (d) show the master curves of the dependence of the storage modulus $G'(\omega)$ and loss modulus $G''(\omega)$ on the angular frequency ω at the reference temperature of 131 $^\circ$C, which suggests the

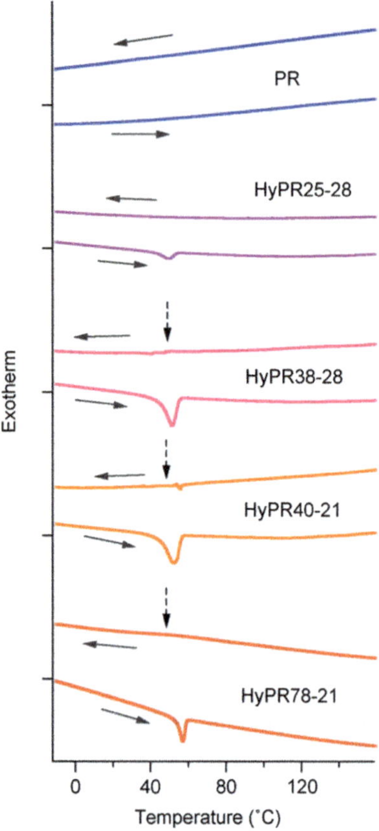

Figure 5.3 Differential scanning calorimetry (DSC) traces (scanning rate: 20 K min^{-1}) of polyrotaxane (PR) and hydroxypropylated polyrotaxane (HyPR) samples. Solid arrows indicate the direction of the scans, while dashed ones point toward the thermal transition points. The lower temperature limit of the graph is set at $-10\,^\circ$C for good legibility of the DSC profiles. No thermal transition was observed below $-10\,^\circ$C, with the exception of the exothermic peak.[7]
Reprinted with permission from *Macromolecules*. Copyright (2010) American Chemical Society.

time–temperature superposition principle for the complex modulus $G^*(\omega) = G'(\omega) + iG''(\omega)$:

$$G^*(T,\omega) = G^*(T_0, a_T\omega), \tag{5.1}$$

where a_T represents the temperature shift factor that is the ratio of a relaxation time at the reference temperature T_0 to that at the measured temperature T. When the time–temperature superposition principle is satisfied, this transition clearly arises from a kinetic phenomenon, such as the glass–rubber transition, rather than a structural transition like the melting process

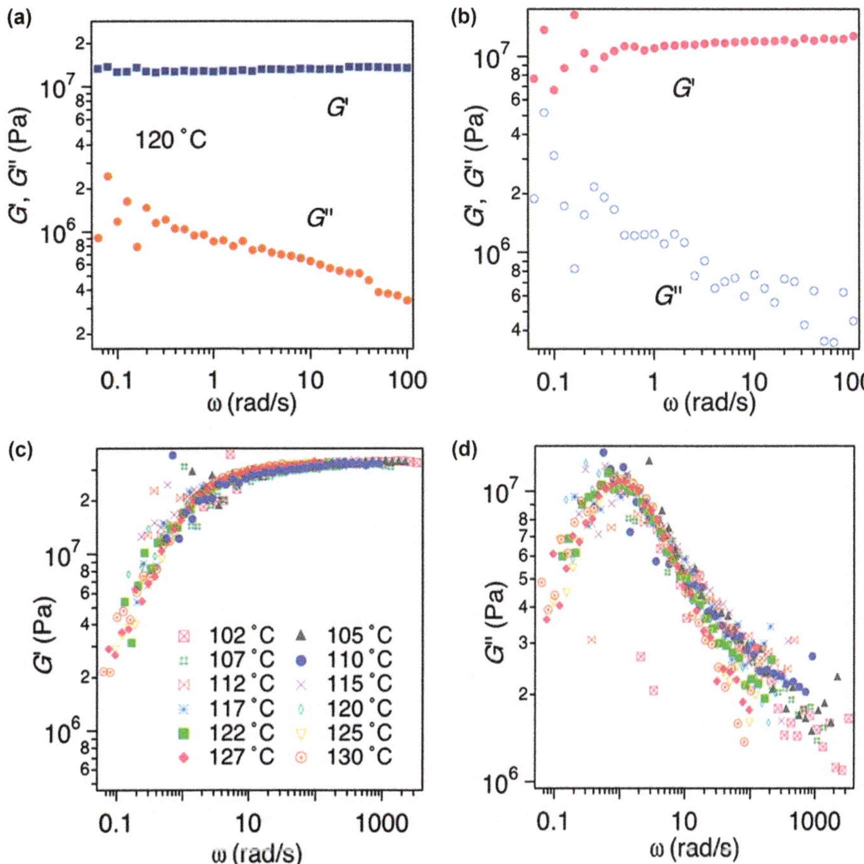

Figure 5.4 Storage and loss moduli of (a) PR at 120 °C, and (b) HyPR25-28 at 120 °C. Master curves of (c) the storage modulus, and (d) the loss modulus of HyPR38-28 are obtained at a reference temperature of 131 °C.[7]
Reprinted with permission from *Macromolecules*. Copyright (2010) American Chemical Society.

or a solid–solid phase change. This kinetic transition corresponds to the thermal transitions observed in the DSC measurements (Figure 5.3).

For the viscoelastic relaxation processes of polyrotaxane derivatives, the temperature dependences of the time-scale shift factors a_T are well described by the empirical Williams–Landel–Ferry (WLF) equation (Figure 5.5), which is often used to analyze relaxation processes related to the dynamic glass transitions of amorphous polymers:[8]

$$\log a_T = \frac{c_1(T - T_0)}{c_2 + (T - T_0)}, \tag{5.2}$$

where c_1 and c_2 are the empirically determined constants, which are related to the free volume of the polymer at the reference temperature T_0. The fitting parameters of c_1 and c_2 are comparable to those of viscoelastic relaxation

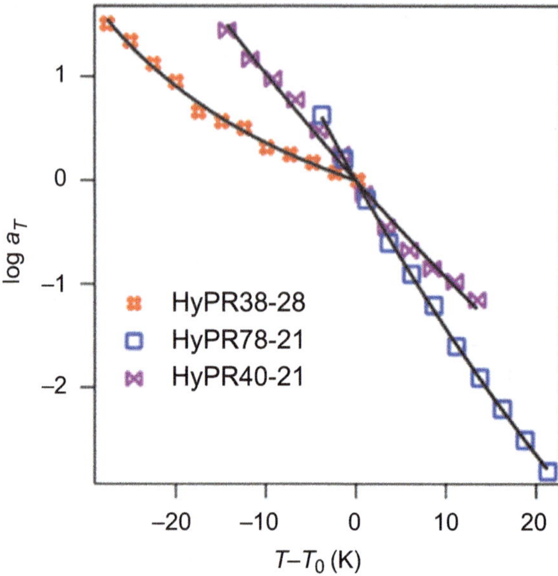

Figure 5.5 Logarithm plot of the temperature dependence of the time-scale shift factor for HyPR38-28, HyPR78-21, and HyPR40-21. The solid curves are the best-fitting ones obtained by eqn (5.2).[7]
Reprinted with permission from *Macromolecules*. Copyright (2010) American Chemical Society.

processes from the glass transitions or dynamic softening transitions of usual polymers.

The WLF-type temperature dependence indicates that a primary dispersion is ascribed to the glass transition, that is, the cooperative segmental motion in hydroxypropylated polyrotaxane. This mechanism is consistent with the thermal transitions commonly observed in the DSC experiments. Because hydroxypropylated cyclodextrins have the largest weight fraction among the components of each hydroxypropylated polyrotaxane, spatial displacement of cyclodextrins should be responsible for the deformation of hydroxypropylated polyrotaxane. Local fluctuations, such as a rotational motion of hydroxypropyl groups on cyclodextrins or a plain rotation of a single cyclodextrin around the cavity axis, cannot largely affect macroscopic strain. Even if poly(ethylene glycol) segments solely fluctuate without displacement of cyclodextrins, hydroxypropylated polyrotaxanes will be deformed slightly. Accordingly, the viscoelastic relaxation process observed in hydroxypropylated polyrotaxanes is attributable to the cooperative motion of several cyclodextrins constrained on the poly(ethylene glycol) backbone together with the micro-Brownian motion of the backbone (Figure 5.6(a)), including the rotational or translational (sliding) motion of cyclodextrins.

The chemical modification of cyclodextrins in solid-state polyrotaxanes significantly changes the interactions among cyclodextrins, such as

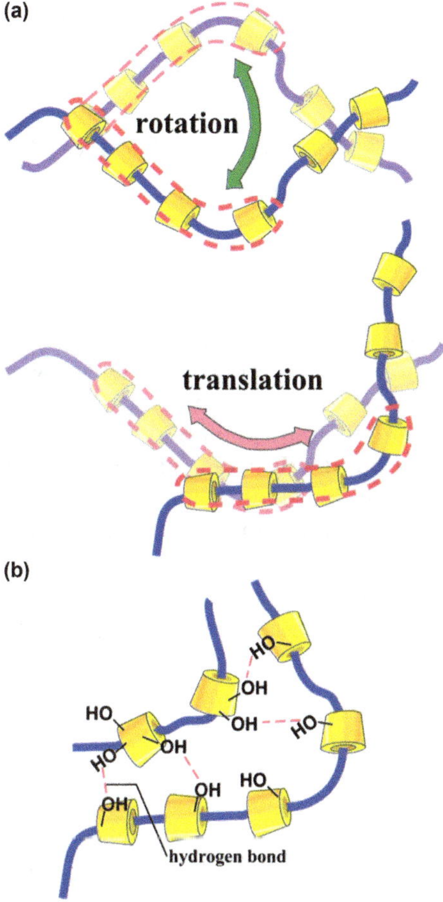

Figure 5.6 (a) Molecular fluctuation of cyclodextrins related to viscoelastic relaxation processes in HyPRs. (b) Hydrogen bonds between residual hydroxyl groups of HyPR25-28.[7]
Reprinted with permission from *Macromolecules*. Copyright (2010) American Chemical Society.

hydrogen bonding, and thereby the mechanical properties of polyrotaxane in the solid state, as well as in solution. Unmodified polyrotaxane maintains the columnar crystal structure of cyclodextrins and shows no viscoelastic relaxation below the decomposition temperature. In contrast, hydroxypropylated polyrotaxanes have disordered packing structures of chemically modified cyclodextrins and viscoelastic relaxation processes corresponding to the dynamic softening of the glassy cyclodextrin domains. The temperature ranges of the viscoelastic relaxations of hydroxypropylated polyrotaxanes depend on the modification ratio: the weaker the molecular interaction among the cyclodextrins becomes due to the steric hindrance by bulky hydroxypropyl groups, the lower the viscoelastic relaxation temperature is. The relaxation processes exhibit WLF-type temperature dependences, which are ascribed to the cooperative rotational or translational motions of several cyclodextrins constrained on the poly(ethylene glycol) backbone together with the backbone string.

5.2 Structure and Dynamics of Polyrotaxane-based Sliding Graft Copolymers

As another important modification of polyrotaxanes, origomers or short polymers can be grafted onto cyclodextrins in polyrotaxanes. Then, these side chains can slide along and rotate around the backbone together with the cyclic molecules, and are called "sliding graft copolymers" (SGCs).[5] Since SGCs have such intramolecular mobility, they may show unique responses to external stimuli such as deformation, temperature, electric field, and so on.

Usual graft copolymers, including comb-like and star polymers, in which the grafting points are covalently fixed, are synthesized by two possible manners: grafting of side chains onto the cyclic molecules ("grafting-to" method), or polymerization of monomers initiated by the functional groups ("grafting-from" method). The same scheme can be applied to SGCs. Sakai *et al.* synthesized SGCs having well-defined side chain lengths by introducing alkyl chains onto cyclodextrins (Figure 5.7), and investigated how the

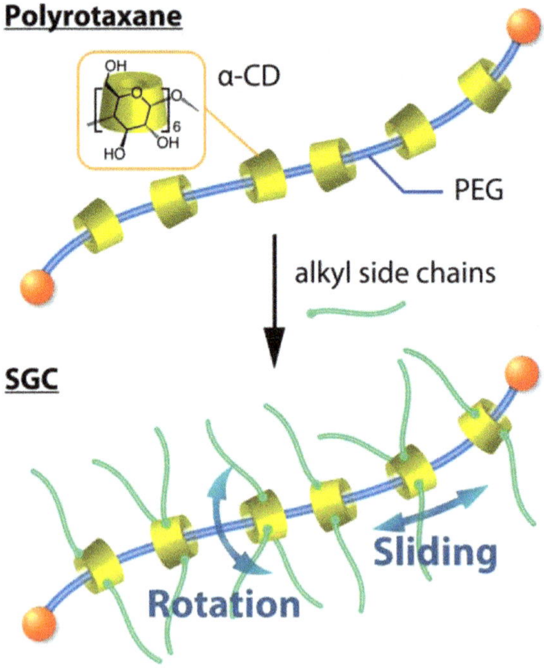

Figure 5.7   Schematic illustrations of a polyrotaxane and a sliding graft copolymer. Reproduced from ref. 9 by permission of The Royal Society of Chemistry.

mobility of the grafting point and the side chain length influenced the structure and dynamics of the SGCs.[9]

Figure 5.8 shows the thermal analysis of SGCs by DSC. An SGC with stearic ($C_{18}H_{37}$) side chains, C18-g-PR, has an endothermic peak around 45 °C on the heating run, which is *ca.* 25 °C lower than that for the melting of stearic acid. This indicates that the C18 side chain in C18-g-PR maintained the crystallinity, forming a partially crystallized structure. The other SGCs and unmodified polyrotaxane show no clear thermal transitions over the measured temperature range. This means that the side chains in these SGCs have non-crystalline structures and no specific structural transitions. The melting point of C18-g-PR is lower than that of stearic acid, which should arise from the suppressed mobility of the grafted side chains and the steric hindrance by cyclodextrins in polyrotaxane. Similarly, the other SGCs show no crystalline structure, while unmodified polyrotaxane holds the hexagonal

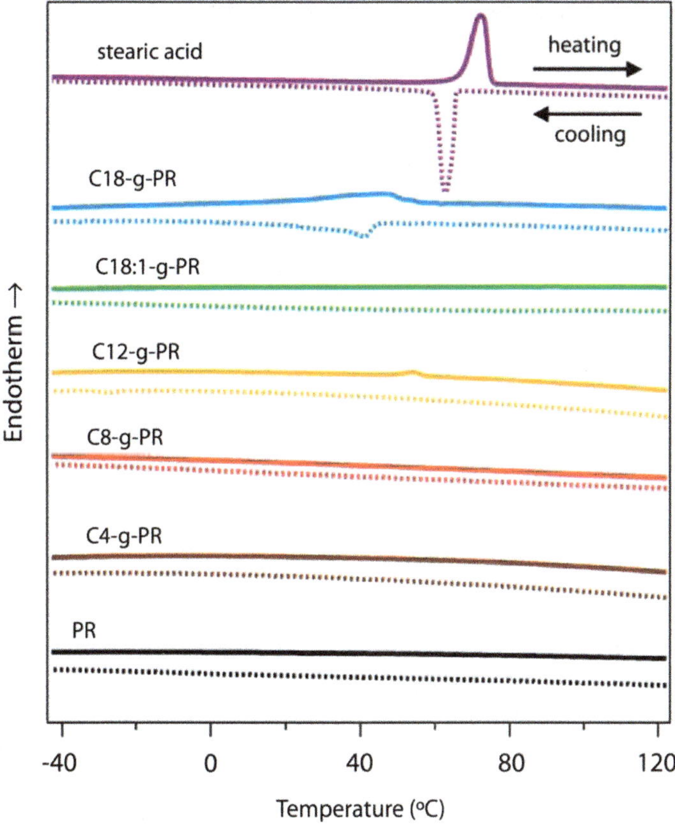

Figure 5.8 DSC charts of stearic acid, sliding graft copolymers, and unmodified polyrotaxane (PR).
Reproduced from ref. 9 by permission of The Royal Society of Chemistry.

crystalline structure of the cyclodextrins at temperatures below the oxidative degradation temperature.[7]

The WAXS profiles of stearic acid (Figure 5.9b(i)) and unmodified poly-rotaxane (Figure 5.9b(vii)) show sharp diffraction peaks from the respective crystal structures. Stearic acids have monoclinic crystals at room tempera-ture,[10,11] while unmodified polyrotaxane forms the hexagonal lattice of α-cyclodextrins in a channel-type crystalline structure by hydrogen bonding between them (Figure 5.2(a)). On the other hand, all the WAXS profiles of the SGC samples (Figure 5.9b(ii)–(vi)) commonly show no sharp crystalline peaks observed in unmodified polyrotaxane but an amorphous halo around $q = 1.4\,\text{Å}^{-1}$. This indicates that alkyl-grafted α-cyclodextrins in SGCs do not form the channel-type crystalline structures. This may result from steric hindrance by the alkyl side chains similar to the disturbance of the inter-molecular hydrogen bonding between the hydroxyl groups of the α-cyclo-dextrins in hydroxypropylated polyrotaxanes.

The WAXS profiles of SGCs (Figure 5.9b(ii)–(vi)) show no sharp diffraction peaks like those observed in stearic acids either, but a broad peak is

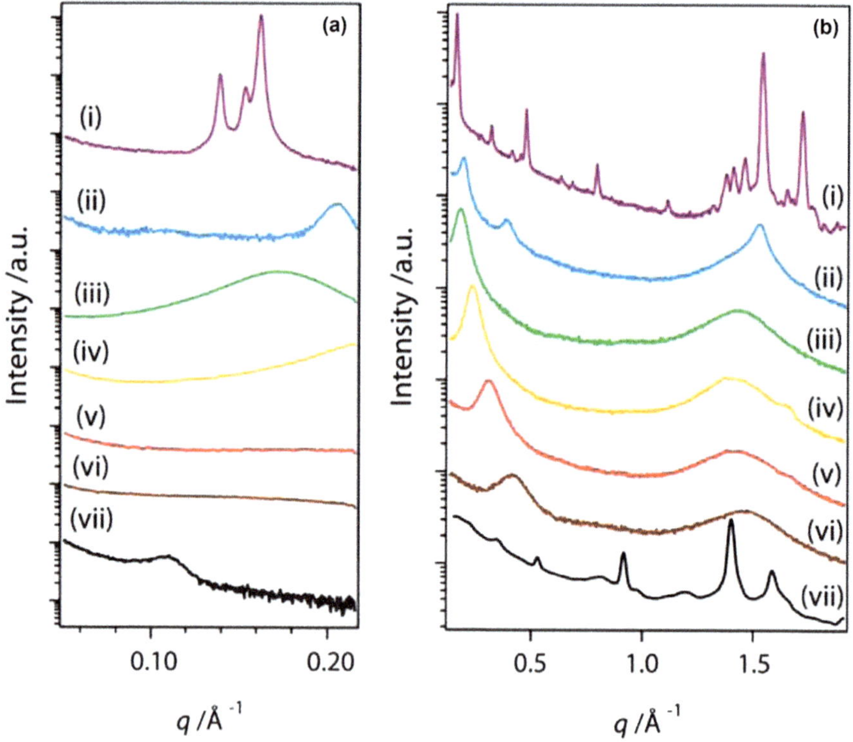

Figure 5.9 (a) SAXS, and (b) WAXS profiles of (i) stearic acid, (ii) C18-g-PR, (iii) C18:1-g-PR, (iv) C12-g-PR, (v) C8-g-PR, (vi) C4-g-PR, and (vii) unmodified PR. Reproduced from ref. 9 by permission of The Royal Society of Chemistry.

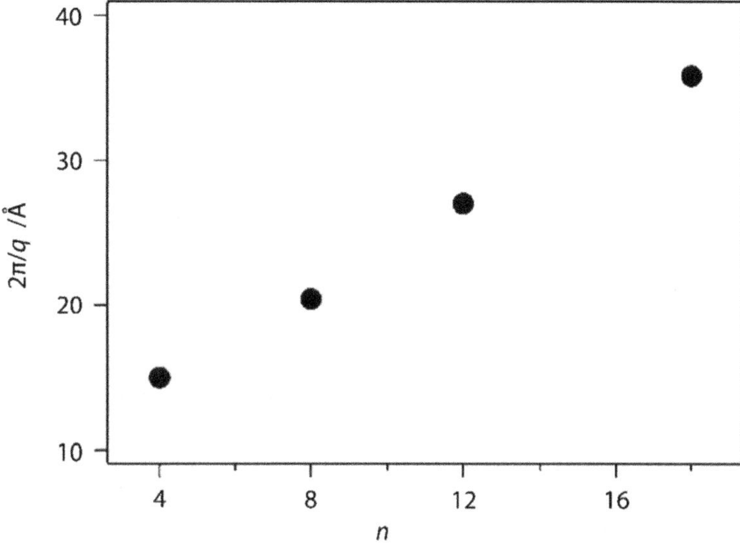

Figure 5.10   Correlation length, $\xi = 2\pi/q$, corresponding to the peak of 0.2–0.4 Å^{-1} in the WAXS profiles of SGCs with non-crystalline side chains plotted against the number of carbon atoms in a side chain, n.[9]
Reproduced from ref. 9 by permission of The Royal Society of Chemistry.

observed in each of the SAXS and WAXS profiles in the range of 0.2–0.4 Å^{-1}. Interestingly, the peak positions shifted to lower q values with increasing alkyl side chain length. This suggests that these peaks are ascribed to the periodic structure related to the side chain length. Figure 5.10 shows the dependence of the correlation length, $\xi = 2\pi/q$, on the number of carbon atoms in the side chains n, where ξ increases linearly with n. Since ξ almost equals the sum of the diameters of the α-cyclodextrins (14.5 Å) and the lengths of the alkyl side chains in all-*trans* conformations, the X-ray scattering profiles suggest the structure of SGCs with non-crystalline side chains and that polyrotaxanes of SGCs are distributed regularly at an average distance of an alkyl side chain length, and the side chains of adjacent SGCs somewhat interdigitate with each other. In comb-like polymers with crystalline side chains, such as poly(n-octadecyl methacrylate), a structural model of interdigitating or end-to-end packing has been proposed. The peak at $q = 0.2$ Å^{-1} (31 Å), observed in both SAXS and WAXS profiles, corresponds to the sum of the α-cyclodextrin radius and C18 chain length, which indicates that the side chains in C18-g-PR form an interdigitating packing structure, as schematically shown in Figure 5.11. The WAXS profile of C18-g-PR shows some small peaks at $q = 0.20$, 0.38 Å^{-1}, and 1.52 Å^{-1}, as well as the amorphous halo commonly seen in other SGCs. The peak at $q = 1.52$ Å^{-1} corresponds to the 110 reflection for stearic acid. This indicates that the C18 side chain of C18-g-PR is slightly crystallized, which is supported by

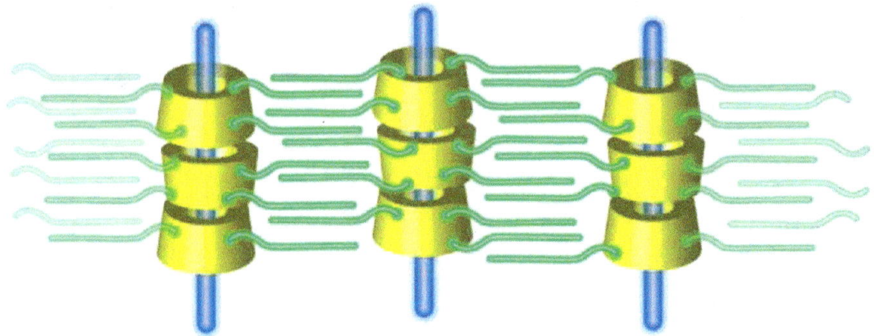

Figure 5.11   Schematic illustration of C18-g-PR structure.
Reproduced from ref. 9 by permission of The Royal Society of Chemistry.

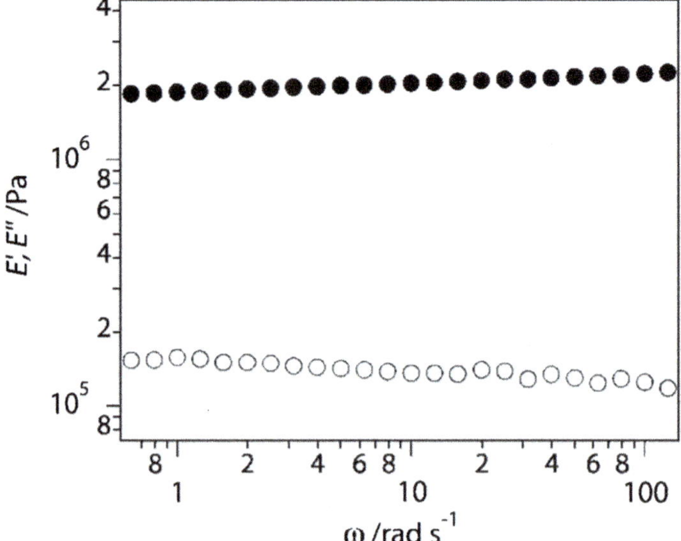

Figure 5.12   Storage (filled circles) and loss (open circles) modulus of C18-g-PR at 20 °C.
Reproduced from ref. 9 by permission of The Royal Society of Chemistry.

the DSC result of C18-g-PR (Figure 5.8), showing a melting behavior of the crystals.

Figure 5.12 shows $E'(\omega)$ and $E''(\omega)$ for SGCs having crystalline C18 side chains, measured at 20 °C, where no relaxation process is observed. This tendency, independent of temperature in the range of 77 to 27 °C, is similar to that of unmodified polyrotaxane, where the crystalline structure or hydrogen bonds of α-cyclodextrins suppresses the molecular fluctuation of α-cyclodextrins. Accordingly, no relaxation process for C18-g-PR indicates

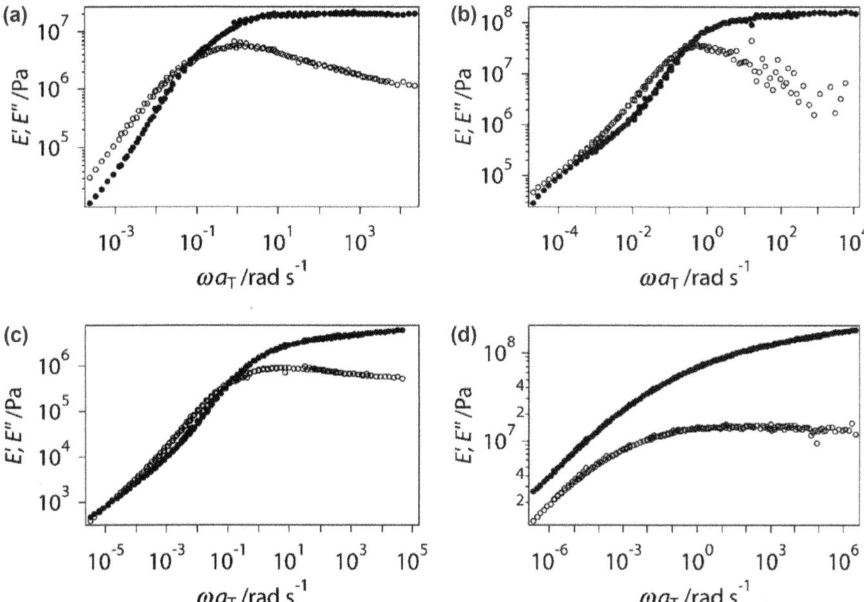

Figure 5.13 Master curves of the storage (E'; filled circles) and loss (E''; open circles) modulus arranged with each reference temperature T_0. (a) C4-g-PR ($T_0 = 50\,^{\circ}\mathrm{C}$), (b) C8-g-PR ($T_0 = 10\,^{\circ}\mathrm{C}$), (c) C12-g-PR ($T_0 = 50\,^{\circ}\mathrm{C}$), and (d) C18:1-g-PR ($T_0 = 60\,^{\circ}\mathrm{C}$).
Reproduced from ref. 9 by permission of The Royal Society of Chemistry.

that the side chain crystallization prevents the molecular fluctuation of α-cyclodextrins substantially.

Figure 5.13 shows the master curves of the other SGCs with non-crystalline side chains based on the time–temperature superposition principle. The viscoelastic spectra at different temperatures can be superimposed by shifting with a shift factor a_T along the logarithmic angular frequencies. Figure 5.14 shows the Arrhenius plot of the shift factor a_T, which gives us the activation energy ΔH.

Clark *et al.* reported the viscoelastic profiles poly(α-olefin) having alkyl side chains of various lengths.[12] The temperature T_α characteristic of the cooperative relaxation mode decreases with increasing side chain length for $n \leq 7$ owing to the plasticizing effect of the side chains. This is because the growth of free volume with increasing side chain length reduces the steric hindrance for segmental motion. In contrast, T_α increases again with n for $n > 7$ since the strengthening interaction between longer side chains should saturate the mobility of side chains. Similarly, ΔH for SGCs also shows a minimum at C8, and increases with the side chain length. Consequently, the relaxation mode of SGCs with non-crystalline side chains is ascribed to the mobility of the side chains and to the cooperative motion of backbone polymer segments and cyclodextrins with side chains.

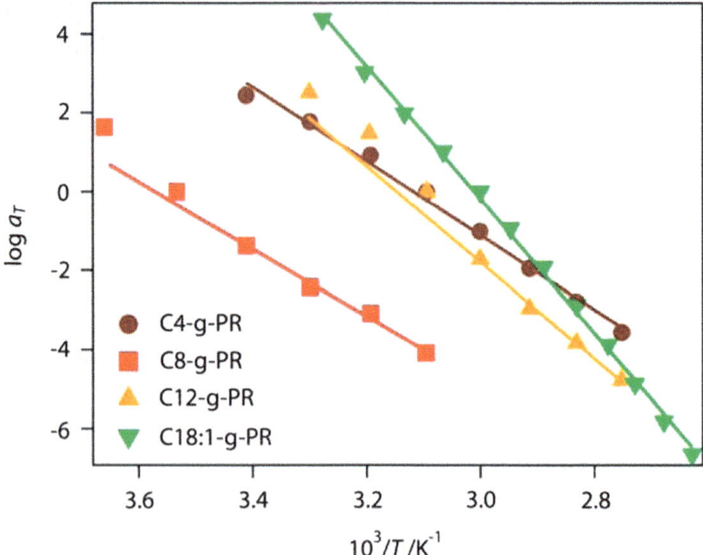

Figure 5.14   Arrhenius plot of the shift factor of C4-g-PR (circles), C8-g-PR (squares), C12-g-PR (triangles), and C18:1-g-PR (inverted triangles). Data are fitted in the lower temperature region (solid lines).
Reproduced from ref. 9 by permission of The Royal Society of Chemistry.

5.3   Orientational Motion in Mesogenic Polyrotaxane and Local Mode Relaxations of Polymer Segments in the Solid State

Kidowaki *et al.* reported mesogenic polyrotaxane by modifying hydroxyl groups of α-cyclodextrins in polyrotaxane with mesogenic groups.[4] Then, mesogens could slide along and rotate around a backbone polymer string, as shown in Figure 5.15. Such a liquid crystalline polyrotaxane built by mesogen-substituted cyclodextrins and poly(ethylene glycol) provides new liquid crystalline polymers with highly mobile mesogenic side chains, in comparison to conventional ones having covalently bonded mesogens.

Figure 5.16 shows DSC thermograms of liquid crystalline polyrotaxane (CB5PR) modified with cyanobiphenyl groups as a side chain. On heating, an endothermic peak appears at 136 °C, while two transitions are seen at 129 and 70 °C upon cooling. The transition at 70 °C is probably ascribed to a glass transition since the transition curve is dependent on the thermal history of the sample. The transition at 129 °C is not seen in unmodified polyrotaxane, which suggests that it is due to a mesomorphic-to-isotropic transition by the modification with mesogenic side chains. As shown in a polarizing optical microphotograph of CB5PR at 100 °C (Figure 5.17), a schlieren-like texture, suggesting a nematic or smectic phase, is observed.

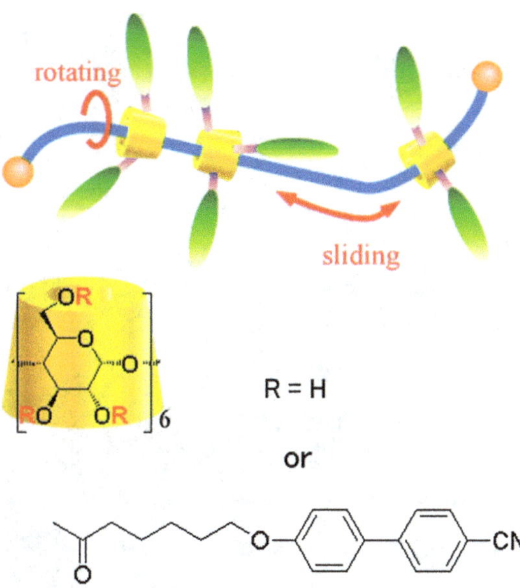

Figure 5.15 Schematic structure of mesogen-substituted polyrotaxane.[4] Reprinted with permission from *Macromolecules*. Copyright (2007) American Chemical Society.

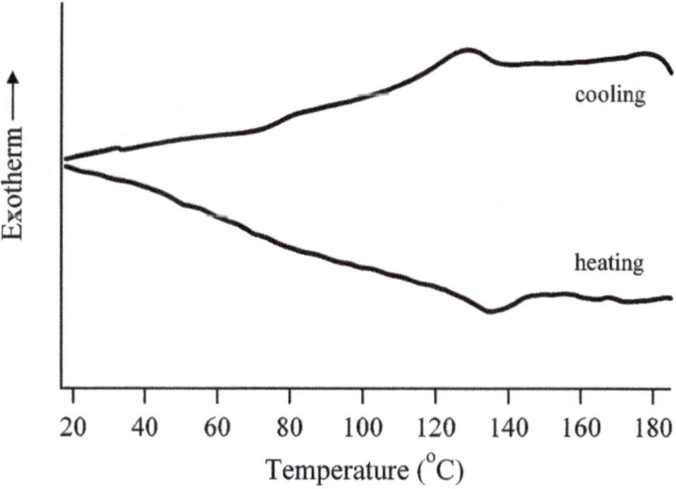

Figure 5.16 DSC thermograms of CB5PR on heating and cooling at 10 °C min^{-1}.[4] Reprinted with permission from *Macromolecules*. Copyright (2007) American Chemical Society.

As temperature decreases, the view field gradually gets brighter but the texture is observed down to room temperature. This suggests that CB5PR forms a liquid crystalline glass state at room temperature.

Figure 5.17   Polarizing optical micrograph of CB5PR at 100 °C.[4]
Reprinted with permission from *Macromolecules*. Copyright (2007) American Chemical Society.

Figure 5.18 shows X-ray diffraction (XRD) patterns of CB5PR at 25, 80, and 180 °C. Two diffraction peaks are observed at $2\theta = 2.9°$ (30 Å) and 20.0° (4.4 Å) at 25 and 80 °C, respectively, while the small-angle peak disappears at 180 °C. Liquid crystalline polymers exhibit the XRD pattern of a diffuse halo in the high-angle region and a small peak in the low-angle one in common. Accordingly, CB5PR can be regarded as a liquid crystalline polymer with movable mesogenic side chains, and as having a nematic mesophase at temperatures below 130 °C.

Inomata *et al.* investigated the dynamics of CB5PR, comparing them to those of unmodified and hydroxypropylated polyrotaxanes by dielectric and viscoelastic relaxation spectroscopy.[13] As shown in Figure 5.19, the dielectric permittivity ε' and loss ε'' of the δ-relaxation process are observed in the annealed CB5PR. The isothermal ε'' curves have a single and distinct maximum at a relaxation frequency increasing with temperature. The maximum frequency in ε'' is clearly seen at a temperature even in the isotropic phase (above 397 K). The shape parameters are evaluated to be $\alpha = 1$ and $\beta = 0.76$–0.83 by fitting the data with the Havilliak–Negami (HN) equation:

$$\varepsilon^{*}(\omega) = \varepsilon' + i\varepsilon'' = \varepsilon_{\infty} + \frac{\Delta\varepsilon}{\left[1 + (i\omega\tau)^{\beta}\right]^{\alpha}} + \frac{\sigma}{(i\omega)^{\gamma}}, \qquad (5.3)$$

which corresponds to the Cole–Cole-type spectra, where ε^* is the complex dielectric constant, ε_{∞} the high frequency limit, i an imaginary number,

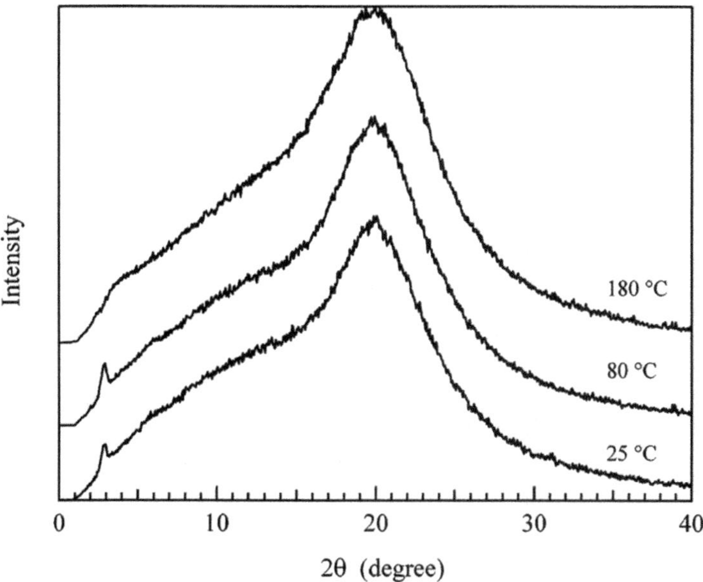

Figure 5.18 X-Ray diffraction (XRD) patterns of CB5PR at 25, 80, and 180 °C.[4] Reprinted with permission from *Macromolecules*. Copyright (2007) American Chemical Society.

σ the conductivity, and γ reflects the effect of the electrode polarization. It is seen that the relaxation strength $\Delta\varepsilon$ and the conductivity σ in a low-frequency range increase drastically with temperature. Figure 5.20 shows the Arrhenius plot of the nominal relaxation time τ. The vertical dashed line exhibits the mesomorphic–isotropic transition point (397 K) of CB5PR. Here, the δ process has a crossover at the nematic–isotropic transition. The activation energy E_A is estimated to be 34 kJ mol^{-1} for the nematic phase and 4.0 kJ mol^{-1} for the isotropic phase.

Since the δ-relaxation process in CB5PR is not seen in unmodified and hydroxypropylated polyrotaxanes, it can be ascribed to the reorientational mode or rotational fluctuation of mesogenic groups around their short axis in CB5PR. The relaxation strength $\Delta\varepsilon$ of the δ-relaxation process in CB5PR increases drastically with temperature, while the activation energy E_A decreases remarkably through the nematic–isotropic transition. The reorientational mode of mesogens in usual liquid crystals shows such temperature dependence of $\Delta\varepsilon$ and E_A in the δ-relaxation. The decrease of E_A is ascribed to the increasing viscosity due to the stronger interaction among mesogens. On the other hand, E_A of the δ process in CB5PR is considerably less than that of usual liquid crystalline polymers ($E_A > 100$ kJ mol^{-1} for the mesomorphic phase, and $E_A > 30$ kJ mol^{-1} for the isotropic phase), although the backbone poly(ethylene glycol) in CB5PR should be more rigid than a usual uncovered one.

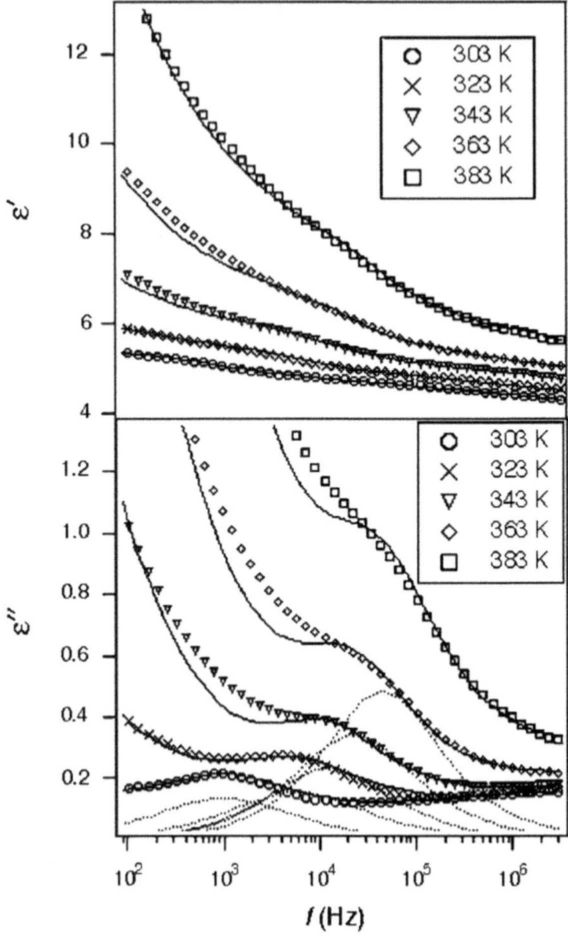

Figure 5.19 Dielectric permittivity (upper) and loss (lower) spectra of the δ process in annealed CB5PR at different temperatures. Solid curves represent the best-fitting results obtained using eqn (5.3).
Reproduced from ref. 13 by permission of The Royal Society of Chemistry.

What is more important is that the δ process in CB5PR is observed at a temperature lower than the glass transition temperature T_g of the poly-rotaxane backbone, where the backbone is in the quasi-glassy state and the cooperative motion of multiple cyclodextrins is frozen.[7] In contrast, usual liquid crystalline polymers show the δ-relaxation strongly coupled with the glass transition of the main chain: the α process of the Vogel–Fulcher–Tammann (VFT)-type temperature dependence.[8] Mesogenic side chains in CB5PR show a mobility higher than that in usual liquid crystalline polymers, even in the glass state of the backbone of polyrotaxane. This is probably because mesogens are bound to cyclodextrins *via* pentyl spacers

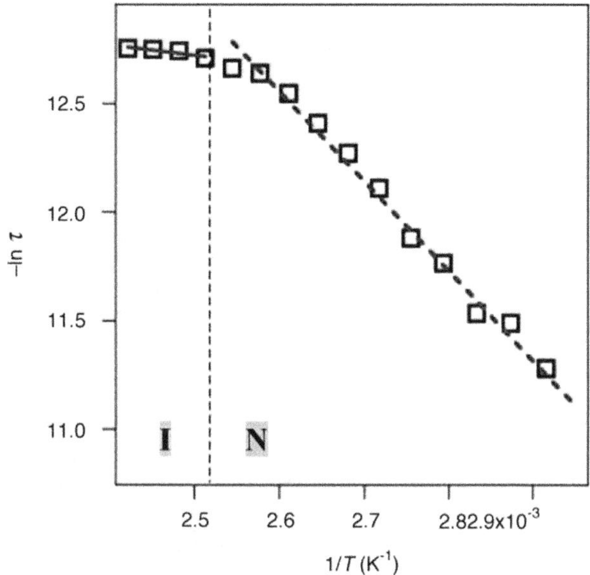

Figure 5.20 Arrhenius plot of the relaxation times for the δ-relaxation process of CB5PR. The vertical dashed line represents the nematic (N)–isotropic (I) transition point of CB5PR.
Reproduced from ref. 13 by permission of The Royal Society of Chemistry.

and are not fixed directly to the backbone poly(ethylene glycol). Cyclodextrins in polyrotaxane can rotate freely around the backbone, which is highly likely to assist the reorientational motion of mesogens in CB5PR and reduce the activation energy considerably.

Figure 5.21 shows the temperature dependence of the dielectric spectra of annealed CB5PR at lower temperature. A broad relaxation mode called the β process is clearly seen in a frequency range of 102–106 Hz at a temperature from 173 K to 213 K, while another mode, the β′ process, is observed in a frequency range (100–104 Hz) lower than that of the β process. Unmodified and hydroxypropylated polyrotaxanes, and poly(ethylene glycol) also show dielectric relaxation in similar temperature ranges. Figure 5.22 exhibits the Arrhenius plots of the relaxation times for all the samples, and the activation energies E_A are summarized in Table 5.1. The activation energy E_A of the β process for CB5PR is almost equal to that of PEG, and slightly less than those for unmodified and hydroxypropylated polyrotaxanes. This means that the β processes for CB5PR, unmodified and hydroxypropylated polyrotaxanes, and poly(ethylene glycol) arise from the same molecular mechanism: the local mode dispersion such as the localized vibrational, torsional, and rotational motions of poly(ethylene glycol) segments in the sub-T_g temperature region. The molecular segment localized within one or two monomer units should fluctuate inside the cyclodextrin cavity of CB5PR, and unmodified and hydroxypropylated polyrotaxanes. The mesogenic interaction between

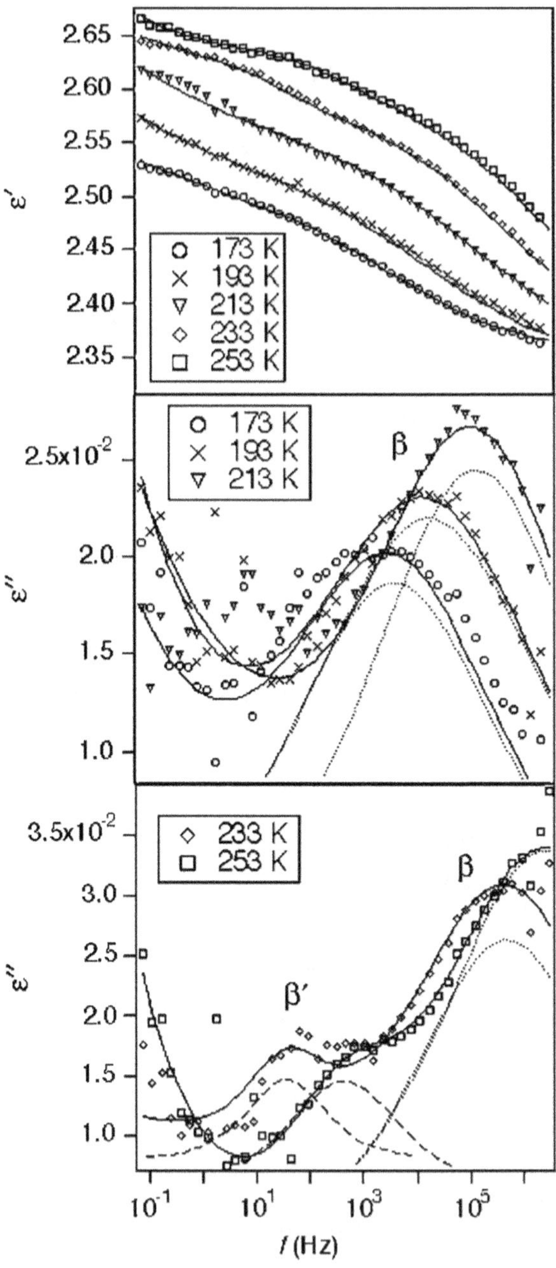

Figure 5.21 Dielectric permittivity spectra at 173–253 K (top), and loss spectra at 173–213 K (middle), and at 233 K and 253 K (bottom) of CB5PR. Solid curves represent the best-fitting results obtained using eqn (5.3). Dotted curves represent the HN terms [second term in eqn (5.3)] of the fitting results for the β process. At 233 K and 253 K, the β′-loss peak (dashed lines) emerges on the low-frequency side of the β-loss peak.
Reproduced from ref. 13 by permission of The Royal Society of Chemistry.

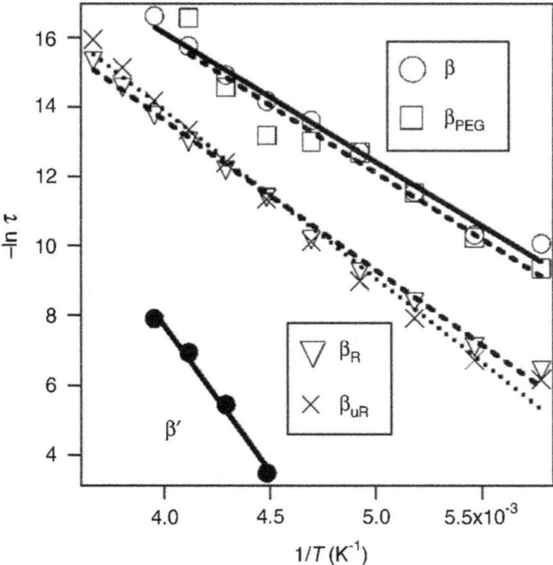

Figure 5.22 Arrhenius plots of the relaxation times for the β- and β'-relaxation processes of CB5PR (β), HyPR (β_R), PEG35 (β_{PEG}), and PR (β_{uR}), where the solid line represents the best fitted ones for β and β' processes, the dashed ones for β_R and β_{PEG}, and the dotted one for β_{uR}.
Reproduced from ref. 13 by permission of The Royal Society of Chemistry.

Table 5.1 Temperature range where each relaxation process occurred in the frequency range of 10^{-1} to 10^6 Hz, Havriliak–Negami fitting parameters in eqn (5.3), and the activation energy E_A for the relaxation processes of CB5PR, HyPR, PR, and PEG. Reproduced from ref. 13 by permission of The Royal Society of Chemistry.

Relaxation process	Temperature range/K	$\Delta\varepsilon$	E_A/kJ mol^{-1}
CB5PR δ	303–413	0.38–2.5	34(N), 4.0(I)
CB5PR β'	223–253	0.020–0.038	70
CB5PR β	173–253	0.14–0.26	31
HyPR β_R	173–273	0.95–1.4	36
PR β_{uR}	173–273	1.62–3.17	40
PEG β_{PEG}	173–243	0.019–0.069	31

cyanobiphenyl groups is not responsible for the local mode dispersion. This means that the mesogenic groups and the backbone poly(ethylene glycol) independently fluctuate and are decoupled with each other in CB5PR since they are not directly connected to each other by a covalent bond.

Cyclodextrins may also affect dielectric relaxation processes because of the large electric dipole moment in the direction of the cyclodextrin cylindrical axis. However, the symmetric rotational mode of cyclodextrin around the backbone poly(ethylene glycol) has no contribution to the dielectric relaxation since the rotation around the cylindrical axis does not affect the direction of the dipole moment of cyclodextrin. This suggests that mesogens with large dipole moments in CB5PR are asymmetrically bonded to cyclodextrins, and thereby contribute to the δ-relaxation. On the other hand, the sliding mode of cyclodextrins in polyrotaxanes is not detected in the measured temperature and frequency ranges. This suggests that the sliding fluctuation of cyclodextrins is quite slow, and requires a considerably higher activation energy than the relaxation processes of polyrotaxanes in the solid state.

5.4 Molecular Dynamics of Polyrotaxanes Investigated by Solid-state Nuclear Magnetic Resonance

Solid-state NMR spectroscopy is a powerful tool to investigate the molecular dynamics of glassy and semi-crystalline polymers, as well as polymer blends and inclusion compounds. Solid-state NMR spectroscopy covers a frequency range of more than 10 orders of magnitude to investigate molecular motion. In particular, detailed dynamic geometric and kinetic parameters are obtained from anisotropic interactions such as ^{2}H quadrupole interactions, chemical shift anisotropy (CSA), and homo- and hetero-nuclear dipole–dipole interactions. Beckham *et al.* reported the molecular dynamics of deuterated poly(ethylene glycol) included in α-cyclodextrin, and the geometry of its motion in the inclusion complex by solid-state ^{2}H NMR.[14] Tonelli *et al.* also investigated the molecular dynamics of polycarbonate,[15] poly-caprolactone,[16] and block copolymers[17,18] in α- or γ-cyclodextrins systematically by a variety of solid-state NMR. For instance, dipolar shift correlation NMR gives the geometry of the local motions of the benzene ring flip on polycarbonate in cyclodextrins. And two-dimensional ^{1}H–^{13}C wideline separation (WISE) provides site-specific dynamics of polycarbonate, both in cyclodextrins and in bulk.

Miyoshi *et al.* investigated the molecular dynamics of both the backbone poly(ethylene glycol) chain and cyclodextrins in unmodified and chemically modified polyrotaxanes systematically using several NMR techniques.[19] Specifically, ^{13}C cross-polarization and magic angle spinning (CP/MAS) NMR provided chemical modification effects on the cyclodextrin side groups from the different chemical shifts. WISE experiments afford a dynamic difference between the cyclodextrins and poly(ethylene glycol), and dynamic heterogeneity in the poly(ethylene glycol) chains. Furthermore, the center-band-only detection of exchange (CODEX) exhibits a direct image of the geometric and kinetic parameters of cyclodextrins with and without chemical modifications.

The effect of the chemical modifications on the polyrotaxane dynamics is well described by WISE, and by CODEX combined with XRD and dynamic mechanical analysis (DMA). The XRD data indicate that unmodified polyrotaxane has a hexagonal crystalline structure. The NMR data indicate that both the overall and side-chain dynamics of cyclodextrins are highly suppressed because of the intermolecular hydrogen bonding among cyclodextrins. Therefore, unmodified polyrotaxane exhibits no relaxation in the temperature range from 303 to 403 K. And WISE clearly shows two components of the poly(ethylene glycol) dynamics at 329 K: the restricted and near-isotropic fluctuation modes assigned to the poly(ethylene glycol) chains covered and uncovered by cyclodextrins, respectively. The covered and uncovered components form domain structures similar to block copolymers (Figure 5.23).

The WISE results indicate that the modification enhances the mobility of cyclodextrins and decreases the restricted components from the poly(ethylene glycol) chains. When the modification ratio is *ca.* 25%, hydrogen bonds suppress cooperative cyclodextrin dynamics significantly, even in disordered states, where most cyclodextrins remain stacked and form the domain structure. On the other hand, the CP/MAS and CODEX results clearly show the overall cyclodextrin motion corresponding to the mechanical relaxation in the higher modification ratio of 78%. Simulations successfully reproduce the CODEX data based on both uniaxial rotational diffusion and random-jump models of cyclodextrins. The two different topologies of cyclodextrin dynamics are due to the spaces available around the cyclodextrins. Limited spaces may allow cyclodextrins to prefer uniaxial rotational diffusion, while complex dynamics, such as rotations, tilting, and sliding motions, would occur even in the solid state if cyclodextrins were randomly dispersed along the backbone poly(ethylene glycol). Although

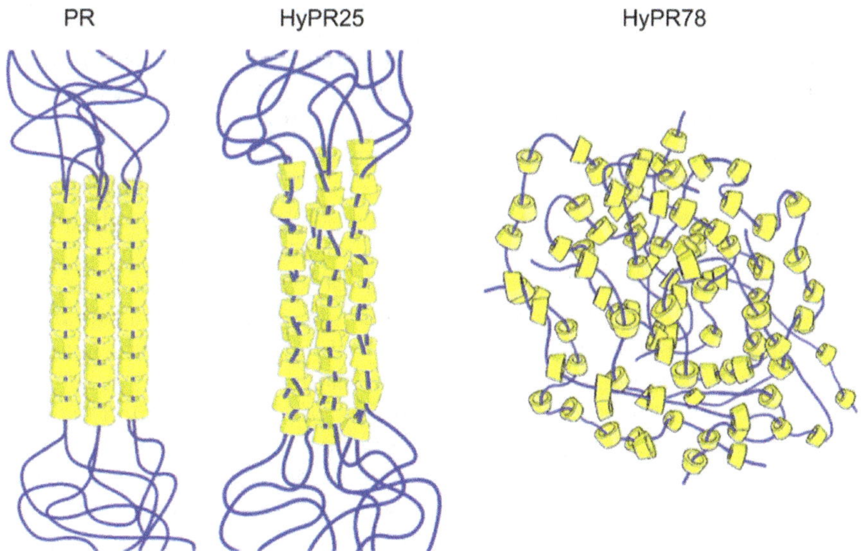

Figure 5.23 Schematic illustrations of the chemical modifications' effects on the structures and dynamics in PR systems.[19]
Reprinted with permission from *Macromolecules*. Copyright (2013) American Chemical Society.

CODEX angular resolution cannot discriminate uniaxial rotational diffusion from the random-jump model sufficiently, WISE clearly indicates that the modification enhances the liquid-like component in the poly(ethylene glycol) chains significantly, and the dynamics of the side chains. These results mean that cyclodextrins without intermolecular hydrogen bonds are randomly distributed along the backbone poly(ethylene glycol) without stacking (Figure 5.23). Consequently, complex dynamics, rotations, tilting, and sliding motions might be preferred, even in the solid state.

The temperature dependence of the correlation time $\langle \tau_c \rangle$ of cyclodextrin backbone dynamics determined by CODEX revealed Arrhenius behavior with a high activation energy 163 kJ mol^{-1}, which is consistent with the viscoelastic result. The high activation energy may arise from complex dynamics, such as translation, rotations, and tilting of cyclodextrins, and/or from cooperative dynamics of cyclodextrins along the backbone poly(ethylene glycol). The intermolecular interactions of hydrogen bonds may yield a higher barrier for apparent cyclodextrin dynamics. On the other hand, WISE indicates that the poly(ethylene glycol) segments perform near-isotropic motions in a fast motional limit ($\langle \tau_c \rangle < 10^{-5}$ s), although cyclodextrins have much slower dynamics, $\langle \tau_c \rangle$ of 0.3 s at 329 K. This suggests that very fast motion of the threaded poly(ethylene glycol) segments does not contribute to cyclodextrin dynamics at the same time scale. Accordingly, very fast poly(ethylene glycol) dynamics is really a local event, and does not move cyclodextrins at the same time scale. In contrast, the slow dynamics of cyclodextrins can move the threaded poly(ethylene glycol) chains performing fast dynamics. This may be another origin for the high activation energy of cyclodextrin dynamics in highly hydroxypropylated polyrotaxanes.

The molecular dynamics of unmodified polyrotaxane has also been investigated in dilute dimethyl sulfoxide (DMSO) solutions by NMR spectroscopy. Beckham and Zhao measured the self-diffusion coefficient of unmodified polyrotaxanes consisting of poly(ethylene glycol) and cyclodextrins with various inclusion ratios of 20–70%, and individual components by diffusion-ordered NMR spectroscopy.[20] Poly(ethylene glycol) has almost the same self-diffusion coefficient as cyclodextrins in unmodified polyrotaxane and is much smaller than those of pure components. The Einstein–Stokes relation gives the hydrodynamic radius of unmodified polyrotaxane and poly(ethylene glycol) in solution, which indicates that various unmodified polyrotaxanes behave as random coil chains in solution independently of the inclusion ratio. From statistical conformational calculations, Tonelli suggested that unmodified polyrotaxane consisted of locally blocky structures of covered and uncovered poly(ethylene glycol) regions even in diluted solution, where several cyclodextrins form a stacking structure by intermolecular hydrogen bonds.[21] This is in contrast to the random distribution of cyclodextrins in highly hydroxypropylated polyrotaxanes. The spatial distributions of cyclodextrins in polyrotaxanes play important roles for the molecular dynamics of poly(ethylene glycol) and the cyclodextrins. Chemical modifications on cyclodextrins can control not only hydrogen bonding among cyclodextrins but also local structures and molecular dynamics of both the cyclodextrins and poly(ethylene glycol) chains in complex polyrotaxanes, which result in physical properties considerably in the solid state.

References

1. M. Kidowaki, C. Zhao, T. Kataoka and K. Ito, *Chem. Commun.*, 2006, 4102.
2. T. Kataoka, M. Kidowaki, C. Zhao, H. Minamikawa, T. Shimizu and K. Ito, *J. Phys. Chem. B*, 2006, **110**, 24377.
3. T. Sakai, H. Murayama, S. Nagano, Y. Takeoka, M. Kidowaki, K. Ito and T. Seki, *Adv. Mater.*, 2007, **19**, 2023.
4. M. Kidowaki, T. Nakajima, J. Araki, A. Inomata, H. Ishibashi and K. Ito, *Macromolecules*, 2007, **40**, 6859.
5. J. Araki, R. Kataoka and K. Ito, *Soft Matter*, 2008, **4**, 245.
6. J. Araki and K. Ito, *Soft Matter*, 2008, **4**, 245.
7. A. Inomata, Y. Sakai, C. Zhao, C. Ruslim, Y. Shinohara, H. Yokoyama, Y. Amemiya and K. Ito, *Macromolecules*, 2010, **43**, 4660.
8. M. L. Williams, R. F. Landel and J. D. J. Ferry, *J. Am. Chem. Soc.*, 1955, 77, 3701.
9. Y. Sakai, R. Gomi, K. Kato, H. Yokoyama and K. Ito, *Soft Matter*, 2013, 9, 1895.
10. E. von Sydow, *Acta Crystallogr.*, 1955, **8**, 557.
11. M. Goto and E. Asada, *Bull. Chem. Soc. Jpn.*, 1978, **51**, 2456.
12. K. J. Clark, A. T. Jones and D. J. H. Sandiford, *Chem. Ind.*, 2010, 1962.
13. A. Inomata, M. Kidowaki, Y. Sakai, H. Yokoyama and K. Ito, *Soft Matter*, 2011, 7, 922.
14. T. E. Girardeau, J. Leisen and H. W. Beckham, *Macromol. Chem. Phys.*, 2005, **206**, 998.
15. Y. Paik, B. Poliks, C. C. Rusa, A. E. Tonelli and J. J. Schaefer, *J. Polym. Sci., Part B: Polym. Phys.*, 2007, **45**, 1271.
16. J. Lu, P. A. Mirau and A. E. Tonelli, *Macromolecules*, 2001, **34**, 3276.
17. F. E. Porbeni, I. D. Shin, X. Shuai, X. Wang, J. I. White, X. Jia and A. E. Tonelli, *J. Polym. Sci., Part B: Polym. Phys.*, 2005, **43**, 2086.
18. J. Lu, P. A. Mirau, I. D. Shin, S. Nojima and A. E. Tonelli, *Macromol. Chem. Phys.*, 2002, **203**, 71.
19. C. Tang, A. Inomata, Y. Sakai, H. Yokoyama, T. Miyoshi and K. Ito, *Macromolecules*, 2013, **46**, 6898.
20. T. Zhao and H. W. Beckham, *Macromolecules*, 2003, **36**, 9859.
21. A. E. Tonelli, *Macromolecules*, 2008, **41**, 4058.

CHAPTER 6

Insulated Molecular Wires based on the Polyrotaxane Structure

6.1 Insulated Molecular Wires

Since polyacetylene was synthesized and the high electric conductivity of doped polyacetylene was reported by Shirakawa, MacDiarmid, Heeger *et al.* in 1977,[1] conjugated conducting polymers have been regarded as promising candidates of the molecular wire. In addition, Friend, Holmes, and co-workers found large electroluminescence in undoped conjugated polymers in 1990.[2] This leads to the application of conjugated conducting polymers as optical molecular devices. These days, conjugated conducting polymers are applied to field effect transistors, light-emitting diodes, photovoltaic cells, and so on. Also they are expected to play important roles in molecular electronics and devices in the future. However, it is widely known that the conducting polymer has some problems upon application as a molecular wire. The strong molecular interactions among conducting polymers prevent a single conducting polymer from being isolated as a molecular wire. Furthermore, a single conducting polymer chain readily forms a coiled conformation to maximize the conformational entropy of the chain, which hinders the formation of π-electron conjugation along the whole chain. Since fascinating electronic and optical properties in a single conducting polymer chain are realized in the *trans* configuration, *i.e.*, the rod-like conformation, we need to isolate a single polymer chain and extend it to the stretched conformation in order to use a conducting polymer as a molecular wire.

Monographs in Supramolecular Chemistry No. 15
Polyrotaxane and Slide-Ring Materials
By Kohzo Ito, Kazuaki Kato and Koichi Mayumi
© Kohzo Ito, Kazuaki Kato and Koichi Mayumi 2016
Published by the Royal Society of Chemistry, www.rsc.org

Polyrotaxane consists of a backbone polymer string and encapsulating rings. If a conjugated conducting polymer and nonconjugated cyclic molecules are used for the backbone string and encapsulating rings, respectively, the polyrotaxane can be regarded as an insulated molecular wire, of which the concept was first mentioned by Maciejewski in the 1970s.[3] Since insulated molecular wires can prevent cross-talk or short-circuits among backbone conjugated polymers, and improve the chemical stability and luminescence efficiency, we may extract the characteristics of the physical properties of an isolated conducting polymer from the insulated molecular wire. Cyclodextrins are nonconjugated cyclic molecules of glucose units, where α-, β-, and γ-cyclodextrins have six, seven, and eight glucose units, respectively. Hence, cyclodextrin-based polyrotaxanes with a conjugated backbone string can form insulated molecular wires. The synthesis, structure, and physical properties of the insulated molecular wire were well summarized in a review paper by Frampton and Anderson.[4]

Anderson *et al.* succeeded in synthesizing cyclodextrins-based insulated molecular wires by Suzuki coupling.[5] They used a diboronic acid and a water-soluble di-iodide, with a small amount of a bulky mono-iodide stopper, to synthesize insulated molecular wires with the backbone of poly(paraphenylene), polyfluorene, poly(4,4'-diphenylenevinylene), and poly(phenylenevinylene), as shown in Figure 6.1. The backbone conjugated polymers include alternating copolymers of ionic and nonionic components, where the encapsulating α- or β-cyclodextrins are located on the nonionic parts separately. The polymerization stoichiometry can control the average degree of polymerization (n) up to $n = 20$. Some insulated molecular

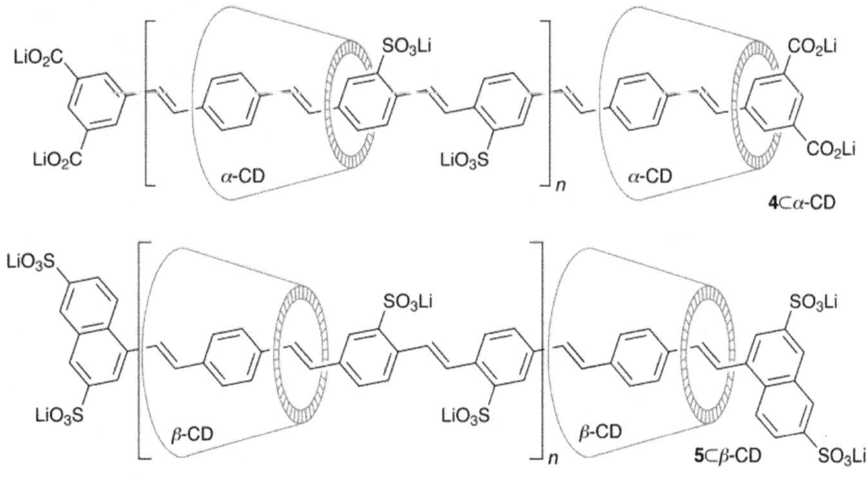

Figure 6.1 Insulated molecular wires of α- or β-cyclodextrins, and poly(phenylenevinylene) (PPV1 $\subset$ β-CD).
Reproduced from ref. 5*d* by permission of The Royal Society of Chemistry.

wires observed by atomic force microscopy (AFM) had almost the same width and height as a cyclodextrin molecule, but longer contour lengths over 100 nm than expected.[4] This suggested the selective adsorption of longer chains onto the mica surface in the spin-coating process. In contrast, pure conducting polymers without cyclodextrins did not show such an image.

Another typical conjugated conducting polymer used for insulated molecular wires is polythiophene and its derivatives. Several groups reported the synthesis of pseudopolyrotaxanes of polythiophene and β-cyclodextrins by polymerization of bithiophene monomers included in β-cyclodextrins. Lagrost and co-workers showed the electropolymerization of polythiophene-based pseudopolyrotaxane in aqueous solution with hydroxyl propylated β-cyclodextrin, which was used rather than unmodified β-cyclodextrin to increase the solubility of the inclusion complex of bithiophene and β-cyclodextrin.[6] Similarly, Harada *et al.* synthesized polythiophene-based pseudopolyrotaxane by oxidizing aqueous solutions of the inclusion complex of bithiophene and β-cyclodextrin with $FeCl_3$.[7] Yamaguchi and coworkers reported the slow kinetics of the inclusion complex formation between polythiophene and β-cyclodextrin, and that a water-soluble pseudopolyrotaxane was formed after stirring polythiophene with β-cyclodextrin for about three weeks in water.[8] Hadziioannou *et al.* succeeded in synthesizing a polythiophene-based polyrotaxane terminated with anthracene by nickel-catalyzed Yamamoto coupling in dimethyl formamide.[9]

Terao reported a new method for synthesizing insulated molecular wires by the polymerization of structurally defined permethylated cyclodextrin-based rotaxane monomers.[10] The insulated molecular wire synthesized had a high covering ratio, rigidity, and high solubility in a variety of organic solvents. As the insulated molecular wire had a rod-like conformation and chiral macrocycles, it showed the cholesteric liquid crystalline phase. The intra-molecular mobility of the backbone conjugated conducting polymer in the insulated molecular wire was measured by both *in situ* flash photolysis time-resolved microwave conductivity (TRMC), and transient absorption spectroscopy (TAS) measurements. The hole mobility along the backbone π-conjugating polymer in the insulated molecular wire was as high as that of amorphous silicon.[11] This also means that the encapsulation of permethylated cyclodextrins prevents π-stacking among conducting polymers in the solid state and block the charge recombination processes, prolonging the lifetime of charged radicals on the backbone. In addition, the insulated molecular wires demonstrated significant fluorescence enhancement, especially in the solid state, which suggested that the encapsulation of the backbone conjugated polymer by permethylated cyclodextrins was effective to increase the fluorescence efficiency of conjugated conducting polymers considerably.

Most conducting polymers without some modification, such as alkyl chains, are insoluble or hardly soluble in solvents. However, such a modification makes it difficult to form the inclusion complex of conducting polymers and encapsulating rings. Polyaniline is an exceptional conjugated polymer and is highly soluble in *N*-methylpyrrolidone, or a mixture of *N*-methyl-2-pyrrolidone and water. This suggests that it is easier to make the inclusion complex of polyaniline and water-soluble cyclodextrins compared to other conducting polymers. In addition, it can be doped by protonation or oxidation, with a drastic color change, as shown in Figure 6.2. The protonation of the emeraldine base to the emeraldine salt increases the conductivity by 10 orders of magnitude up to *ca.* 10^2 S cm^{-1}. As a result, polyaniline is applied to various items such as batteries, sensors, actuators, and electrochromic devices.[12]

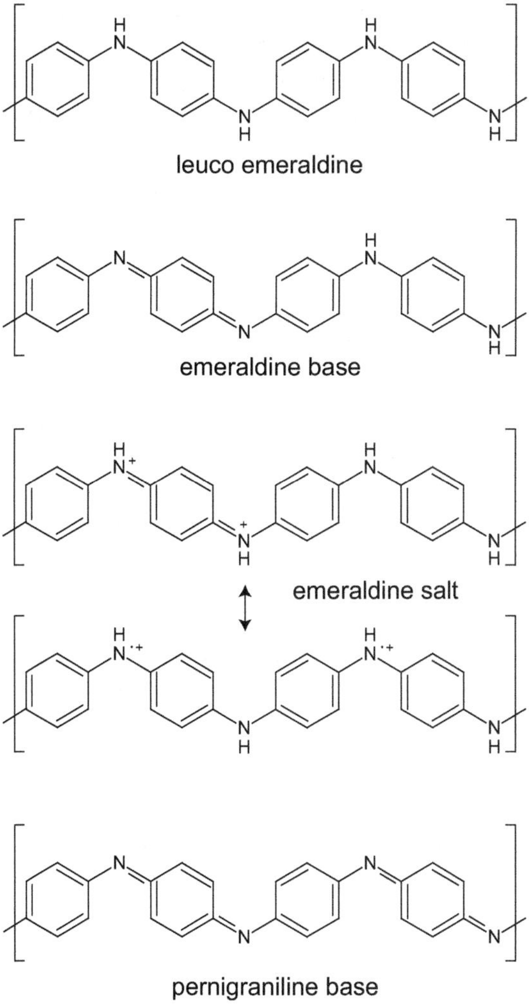

Figure 6.2 The redox and protonation states of polyaniline.

6.2 Polyaniline-based Insulated Molecular Wires

Shimomura and coworkers first synthesized polyaniline-based pseudopolyrotaxane and investigated the structure and properties by frequency-domain electric birefringence (FEB) spectroscopy, scanning tunneling microscopy (STM), and AFM.[13] The FEB technique probes the dynamic Kerr effect based on molecular optical and electric anisotropy in solution. This means that the rod-like molecules with optical and electric anisotropy show large electric birefringence, whereas isotropic coiled polymer chains give no response. In the FEB method, we measure the birefringence signal Δn of the solution when we apply the sinusoidal electric field $E = E_0 \sin\omega t$ of the angular frequency ω, t the time, and the amplitude E_0 to the specimen.[14] If molecules have optical anisotropy and are oriented toward the applied electric field in the solution, the solution demonstrates birefringence, which is defined as the Kerr effect, a second-order nonlinear optical effect. The FEB gives us the frequency-dependent Kerr constant, $K \equiv \dfrac{\Delta n}{E_0^2} = K_{dc} + \mathrm{Re}\left[K_{2\omega}^* \exp(2i\omega t)\right]$, where K_{dc} represents the dc component and $K_{2\omega}^* \left(= K_{2\omega}' - iK_{2\omega}''\right)$ is the complex amplitude of the 2ω component and i an imaginary number. The theoretical analysis indicates that $K_{dc}(= \mathrm{Re}[\psi^*])$ gives the information on the anisotropy and dynamics of the electrical polarizability of the molecule, where ψ^* is the complex amplitude of the Kerr response.[14] And $K_{2\omega}^* = \psi^* / [1 + i(2/3)\omega\tau_r]$ shows the molecular shape through the rotational relaxation time τ_r.[14] What is important is that the FEB response appears only in solutions of rod-like molecules with high electro-optical anisotropy.[15]

A small amount of *N*-methyl-2-pyrrolidone solution of polyaniline was mixed into an aqueous solution of β-cyclodextrin, as shown in Figure 6.3, and cooling the mixture yielded blue precipitation in the solution.[13] Such precipitation did not appear in an aqueous solution of β-cyclodextrin nor *N*-methyl-2-pyrrolidone solution of polyaniline. Next α-cyclodextrin and polyaniline were mixed similarly and cooled down, but no precipitation was observed. This suggests the inclusion complex formation of polyaniline-based pseudopolyrotaxane of β-cyclodextrin and polyaniline.

Figure 6.4 shows typical FEB spectra of the mixed solution of β-cyclodextrin and polyaniline in *N*-methyl-2-pyrrolidone (a) at 255 K, and (b) at 300 K.[13] While the FEB signal cannot be detected in the mixed solution at 300 K, it appears around 275 K and enlarges as temperature T decreases. In contrast, no FEB response was observed in *N*-methyl-2-pyrrolidone solutions of polyaniline only, even at low temperatures down to 250 K. This means that rod-like molecules with electro-optical anisotropy are formed in the mixture of β-cyclodextrin and polyaniline in *N*-methyl-2-pyrrolidone at low temperatures below 275 K, and the anisotropy increases with decreasing T. Since the inclusion complex formation is favorable at lower temperature because of the entropic disadvantage, the experimental results support the inclusion complex formation of a rod-like pseudopolyrotaxane of β-cyclodextrin and polyaniline with electro-optical anisotropy.

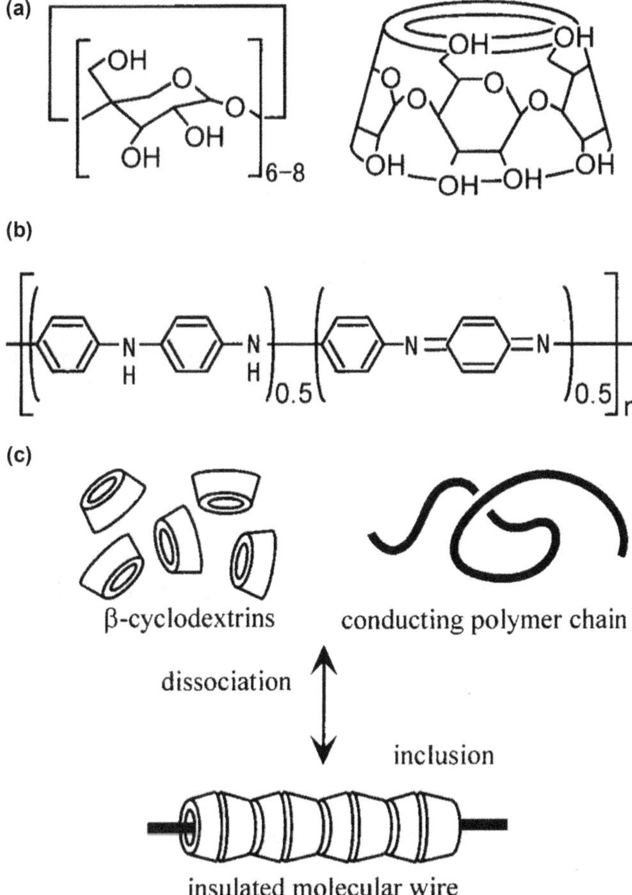

Figure 6.3 Schematic diagrams of (a) cyclodextrins, (b) polyaniline with emeraldine base, and (c) inclusion complex formation of cyclodextrins and a conducting polymer chain: insulated molecular wire.[13]
Reprinted with permission from *Langmuir*. Copyright (1999) American Chemical Society.

The rotational relaxation frequency $f_r \equiv 3/(4\pi\tau_r)$, at which $K''_{2\omega}$ shows a maximum, gives the effective length (L_{eff}) of a rod-like molecule as:[14]

$$f_r = \frac{9k_B T}{2\pi^2 \eta_0 L_{\text{eff}}^3} [\ln(L_{\text{eff}}/d_r) + \gamma_r], \qquad (6.1)$$

where η_0 is the solvent viscosity, d_r is the diameter of the rod-like molecule, γ_r is a correction term for the end effect, and $k_B T$ is the thermal energy. Figure 6.5 shows the temperature dependence of L_{eff}, where no FEB signal was found at temperatures higher than 275 K. As shown in Figure 6.5, L_{eff} is nearly equal to 1.5 μm almost independently of temperature, which is longer than the contour length of polyaniline (0.3 μm) evaluated from the molecular weight.

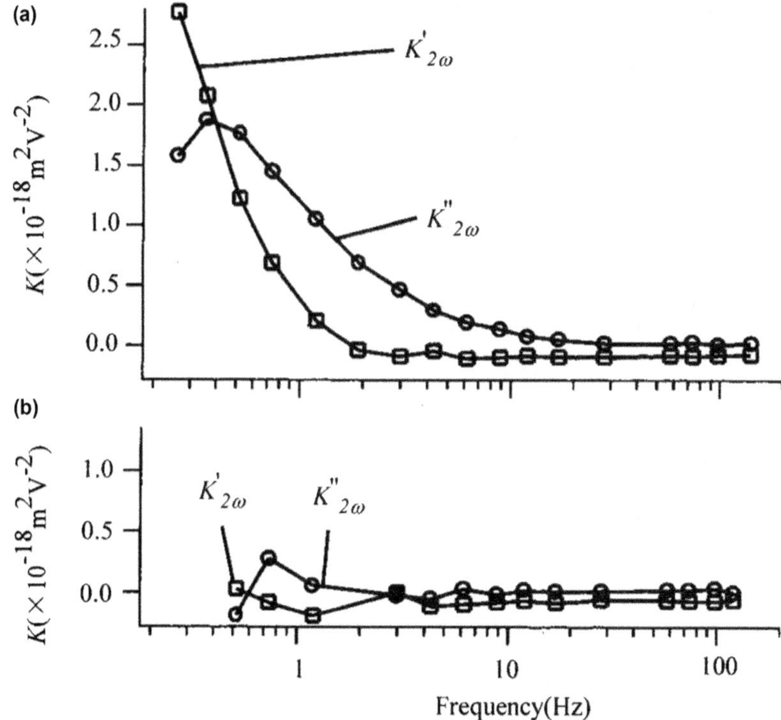

Figure 6.4 Typical experimental results of the FEB spectra in the mixed solution of β-cyclodextrin and polyaniline in *N*-methyl-2-pyrrolidone (a) at 255 K, and (b) at 300 K.[13]
Reprinted with permission from *Langmuir*. Copyright (1999) American Chemical Society.

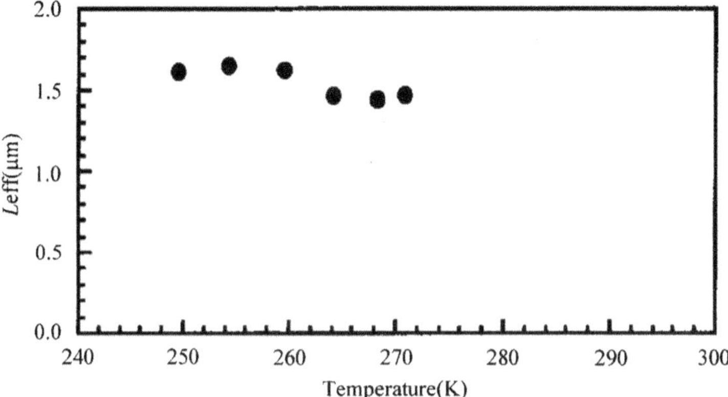

Figure 6.5 The temperature dependence of the effective length L_{eff} calculated from the rotational relaxation time τ_r with eqn (6.1). At temperatures above 275 K, the FEB signal was not detected.[13]
Reprinted with permission from *Langmuir*. Copyright (1999) American Chemical Society.

The image of the polyaniline-based pseudopolyrotaxane was observed by STM, as shown in Figure 6.6, when a drop of the mixed solution cooled at 255 K was spin-coated onto a substrate of highly ordered pyrolytic graphite. The length of the rod-like structure (330 nm) is consistent with the contour length of polyaniline. Figure 6.6(c) indicates that the vertical height of the structure is about 2 nm, which is equal to the outside diameter of β-cyclodextrin. This also supports the inclusion complex formation of β-cyclodextrin and polyaniline.

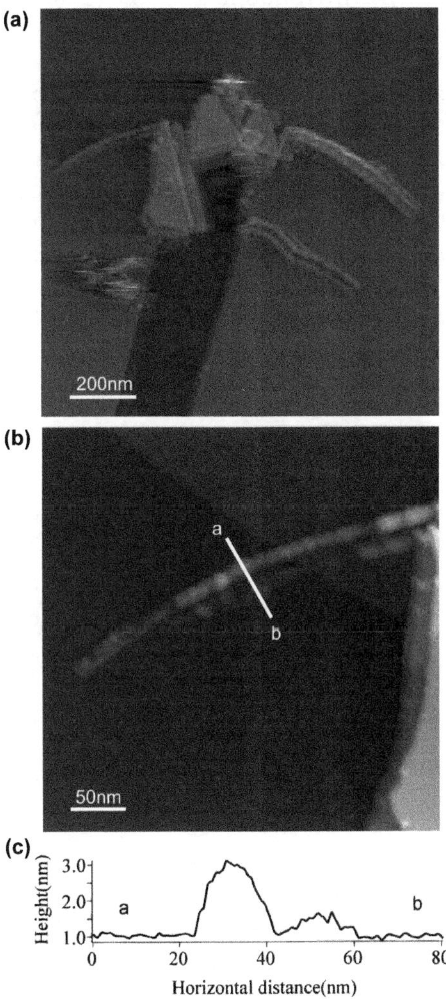

Figure 6.6 STM images, where the fields of view are (a) 1200 nm × 1200 nm, and (b) 350 nm × 350 nm. (c) The height profile of the cross section.[13] Reprinted with permission from *Langmuir*. Copyright (1999) American Chemical Society.

To further confirm the inclusion complex formation, iodine was added to the mixed solution of β-cyclodextrin and polyaniline, or a solution of polyaniline only at 255 K. Iodine oxidizes polyaniline with emeraldine base and changes the color of the polyaniline solution from blue to violet. The color change between the mixture of β-cyclodextrin and polyaniline was compared to the polyaniline solution at 273 K. The polyaniline solution changed color from blue to violet after a few days, while the mixed solution remained blue. As temperature further increased, the mixed solution also changed color from blue to violet with a considerable delay compared to the polyaniline solution. This suggests that polyaniline forms the inclusion complex with β-cyclodextrin in *N*-methyl-2-pyrrolidone at a low temperature and is almost fully covered by encapsulating β-cyclodextrins.[13]

Next, Shimomura *et al.* tried to form another inclusion complex of polyaniline and molecular tubes made from α-cyclodextrins, as shown in Figure 6.7.[16] Harada *et al.* successfully synthesized molecular tubes by cross-linking adjacent α-cyclodextrins in a polyrotaxane with poly(ethylene glycol) as an axis polymer.[17] The molecular tube can form the inclusion complex with linear polymer chains more effectively than cyclodextrins because the entropy loss accompanied with the inclusion complex formation of tubes and strings is generally much less than that of rings and strings, as suggested in Chapter 2. Since a molecular tube of *ca.* 0.45 nm inside diameter

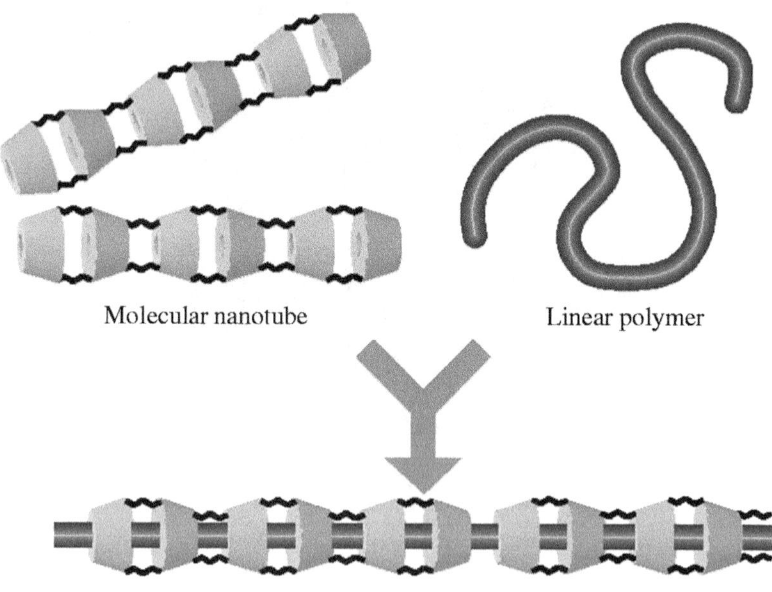

Figure 6.7 Inclusion complex formation between a linear polymer chain and molecular tubes.[16] Used in accordance with the Creative Commons Attribution 3.0 Unported License.

confines polyaniline to a rod-like conformation or all-*trans* configuration (because of the very small cavity), the gauche configuration as a defect of the π-conjugated system should be suppressed on polyaniline.

When Shimomura *et al.* mixed a small amount of blue *N*-methyl-2-pyrrolidone solution of emeraldine base polyaniline into an aqueous solution of the molecular tube at room temperature, blue fibrous precipitation was observed in the solution, which increased as the concentration of molecular nanotube enlarged. In contrast, the precipitation was not seen in a mixture of polyaniline solution without the molecular tube in *N*-methyl-2-pyrrolidone and water. This difference suggests the inclusion complex formation between polyaniline and the molecular tube as an insulated molecular wire. As mentioned before, the inclusion complex formation of polyaniline and β-cyclodextrin occurred only at temperatures lower than 275 K. In contrast, the molecular tube can form an inclusion complex with polyaniline even at room temperature. This indicates that the molecular tube has a stronger inclusion force on polyaniline than β-cyclodextrin.

The *N*-methyl-2-pyrrolidone solution with fibrous precipitation was spin-coated on a mica substrate. As shown in Figure 6.8(a), a rod-like structure of

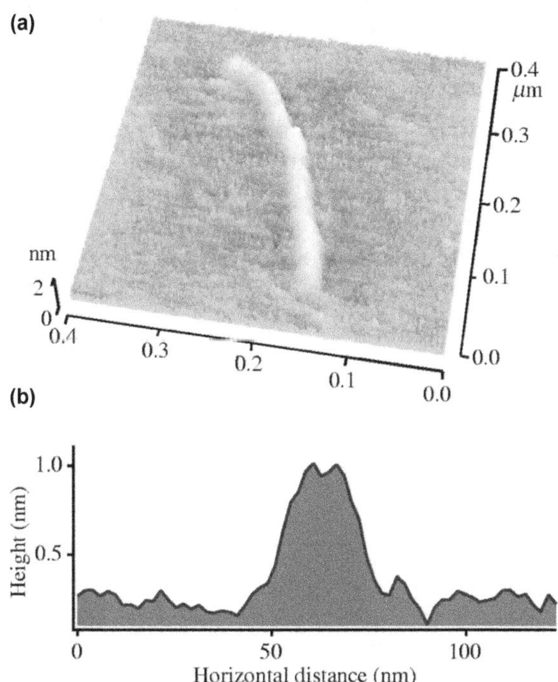

Figure 6.8 (a) AFM topographic image (400 nm × 400 nm) of cleaved the mica substrate on which the mixed solution of polyaniline and the molecular tube was spin-coated. (b) The height profile of the cross section.[16] Used in accordance with the Creative Commons Attribution 3.0 Unported License.

ca. 300 nm in length was observed. The size almost agrees with the contour length of polyaniline evaluated from the averaged molecular weight. Figure 6.8(b) shows a vertical height of the structure of about 1 nm, close to the outside diameter of the molecular tube. Accordingly, this rod-like structure was identified as an inclusion complex between polyaniline and the molecular tube, *i.e.*, the insulated molecular wire. And it has almost uniform height along the whole length. This means that polyaniline is nearly fully covered by the molecular tubes. Incidentally, no rod-like structures were observed on mica substrates prepared without the precipitation from the pure polyaniline solution.

The length histogram for inclusion complexes showed inclusion complexes 300–400 nm in length most frequently (Figure 6.9). The length is consistent with the averaged contour length of polyaniline evaluated from the averaged molecular weight, but the distribution was asymmetric, with longer inclusion complexes than expected from the gel permeation chromatography (GPC) results. Figure 6.10 shows long insulated molecular wires, which were not observed from the mixed solution of polyaniline and β-cyclodextrin. Hence, these long insulated molecular wires may be formed by linkage of some polyaniline chains with molecular tubes.

The inclusion complex formation of polyaniline and β-cyclodextrin needs the excess amount of β-cyclodextrin, typically about 900 times the weight of polyaniline, as well as a low temperature below 270 K. A small amount of the molecular tube, less than the weight of polyaniline, can yield the inclusion complex with polyaniline at room temperature. These results suggest that the molecular tube has a much stronger inclusion interaction with polyaniline than β-cyclodextrin.

Figure 6.11 shows the manipulation of an insulated molecular wire with AFM tips. When the AFM tip was moved from position A to B under contact

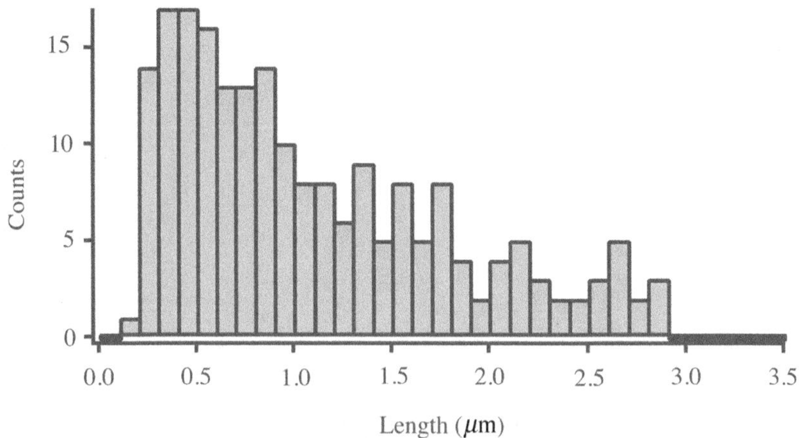

Figure 6.9 Length histogram for insulated molecular wires in 210 AFM images.[16] Used in accordance with the Creative Commons Attribution 3.0 Unported License.

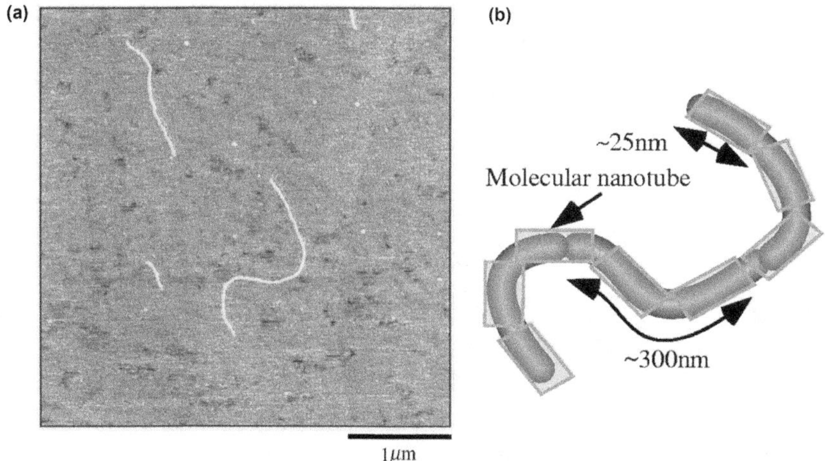

Figure 6.10 (a) AFM topographic image (4 mm × 4 mm) of insulated molecular wires more than 2 µm in length on a cleaved mica substrate. (b) Schematic image of an insulated molecular wire formed by molecular nanotubes linking some polyaniline chains.[16] Used in accordance with the Creative Commons Attribution 3.0 Unported License.

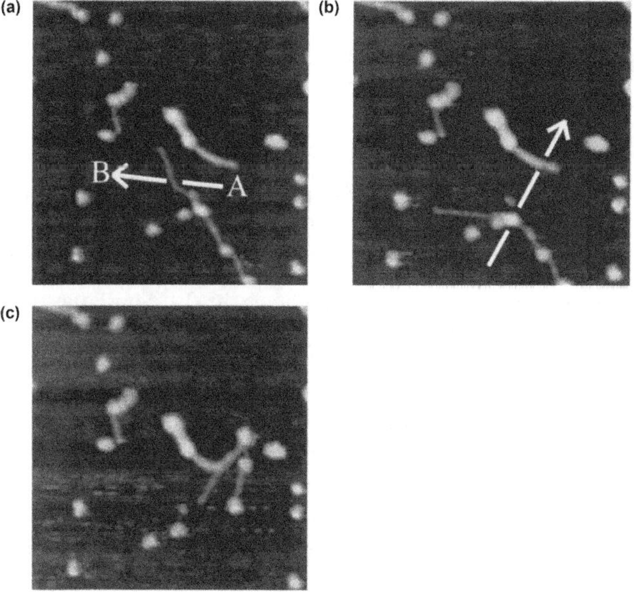

Figure 6.11 AFM images of an insulated molecular wire on a mica substrate (scan size: 620 nm × 620 nm). The sequence depicts the AFM manipulation process. The white arrow represents the movement of the cantilever tip in contact mode. After the tip was moved from position A to B in (a), the image in (b) was obtained, where the insulated molecular wire was moved and bent. Similarly, the subsequent manipulation procedure moved the wire as shown in (c).[18]
Reproduced by permission of The Japan Society of Applied Physics.

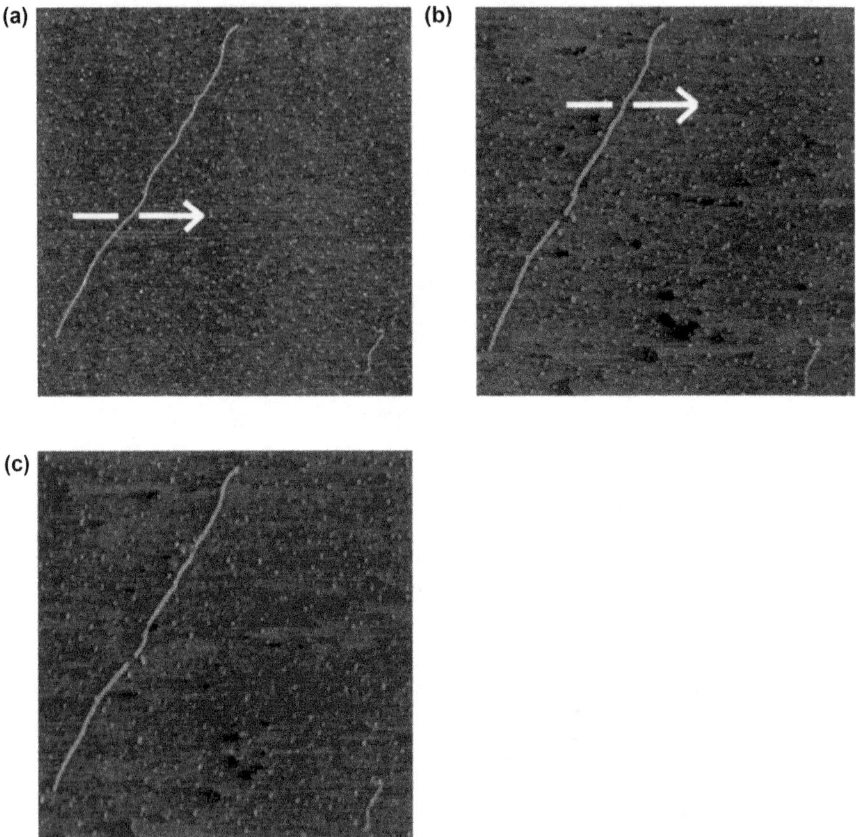

Figure 6.12   AFM images of an insulated molecular wire on a mica substrate (scan size: 2300 nm × 2300 nm). The sequence depicts the AFM manipulation process. The white arrow represents the movement of the cantilever tip in contact mode. After the tip crossed the insulated molecular wire in (a), the wire was cut off as shown in (b). Similarly, the subsequent manipulation cut the wire as shown in (c).[18]
Reproduced by permission of The Japan Society of Applied Physics.

mode control, the insulated molecular wire was bent.[18] A similar subsequent manipulation moved the wire, as shown in Figure 6.11(c). Figure 6.12 demonstrates that an insulated molecular wire was cut off with the tip crossing the molecular wire in contact mode.

Shimomura *et al.* reported the first conductivity measurements on single insulated molecular wires of a polyaniline-based pseudopolyrotaxane.[19] A single insulated molecular wire was placed across platinum narrow-gap electrodes 150 nm in width on a sapphire substrate, as shown in Figure 6.13. The conductance of the emeraldine base or undoped insulated molecular wire was too small to be detected, which indicated that the undoped insulated molecular wire was an insulator. Incidentally, the emeraldine base

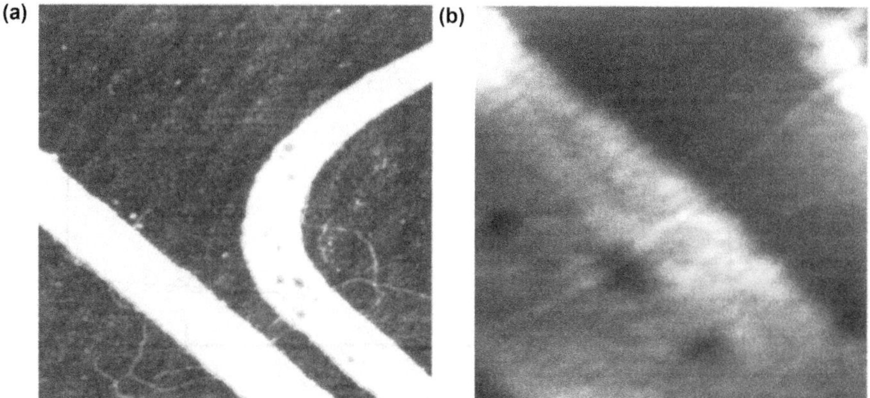

Figure 6.13 AFM topographic image of an insulated molecular wire positioned across Pt narrow-gap (150 nm) electrodes on a sapphire substrate; scan size (a) 2.0 μm × 2.0 μm, and (b) 400 nm × 400 nm.[19]
Reprinted from *Synth. Met.*, **153**, 497–500, Copyright (2005), with permission from Elsevier.

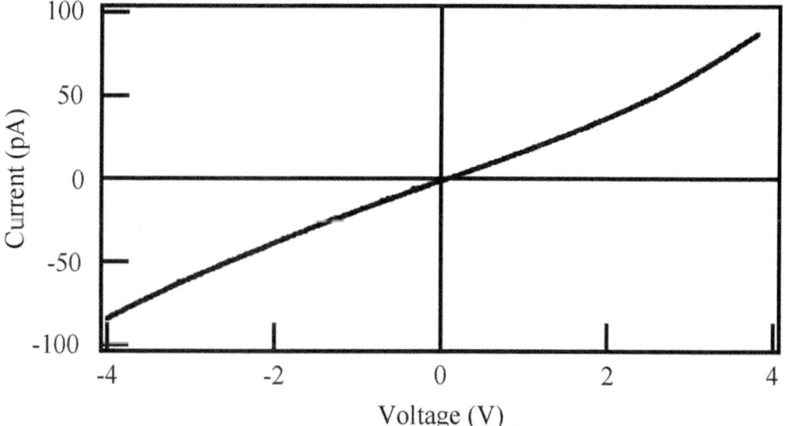

Figure 6.14 *I–V* profiles of the insulated molecular wire (a) before doping, and (b) upon doping with iodine for 16 hours.[19]
Reprinted from *Synth. Met.*, **153**, 497–500, Copyright (2005), with permission from Elsevier.

polyaniline film also shows quite small conductivity of *ca.* 10^{-10} S cm^{-1}. When the insulated molecular wire was doped with iodine for 16 hours, the current–voltage (*I–V*) characteristics changed drastically at 30 °C, as shown in Figure 6.14. The resistance was evaluated to be 47 GΩ from the *I–V* characteristics in Figure 6.14.

To confirm that the conductance arises from the insulated molecular wire, Shimomura *et al.* measured the conductance of the narrow-gap electrode

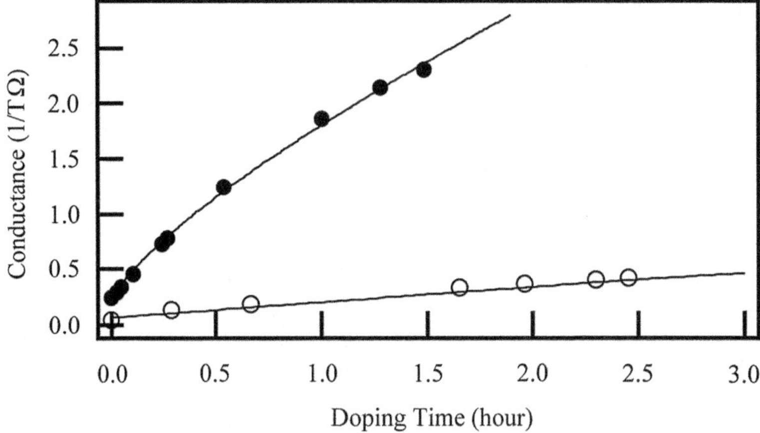

Figure 6.15   Conductance measured between the Pt narrow-gap electrodes as a function of exposure time to iodine vapor. Open circles denote the conductance between them without the insulated molecular wire, while filled ones indicate the conductance in the presence of the insulated molecular wire.[19]
Reprinted from *Synth. Met.*, **153**, 497–500, Copyright (2005), with permission from Elsevier.

both bridged by the insulated molecular wire and in the absence of it. Figure 6.15 shows the dependence of the conductance of both samples on the exposure time of iodine vapor. The conductance of the gap electrodes with the insulated molecular wire is an order of magnitude larger than that without it. This means that the conductance is ascribed to the insulated molecular wire. The conductivity of the single insulated molecular wire was evaluated to be 4×10^{-2} S cm^{-1} from the resistance of 47 GΩ, which was a little smaller than the conductivity of the doped polyaniline film with iodine (10^{-1} S cm^{-1}). If a single conducting polymer chain is fully extended and highly doped, it is expected to show much higher conductivity. The low conductivity of the insulated molecular wire suggests that iodine doping is not enough, since molecular tubes may prevent iodine from contacting directly with polyaniline. The mechanism of charge transport between polyaniline and iodine, or between polyaniline and the platinum electrode through a molecular tube of α-cyclodextrin is not yet fully understood.

6.3   Structural Analysis of Insulated Molecular Wires

Belosludov and coworkers investigated the structural configuration of polyaniline in pseudopolyrotaxane with β-cyclodextrins by the two-layered

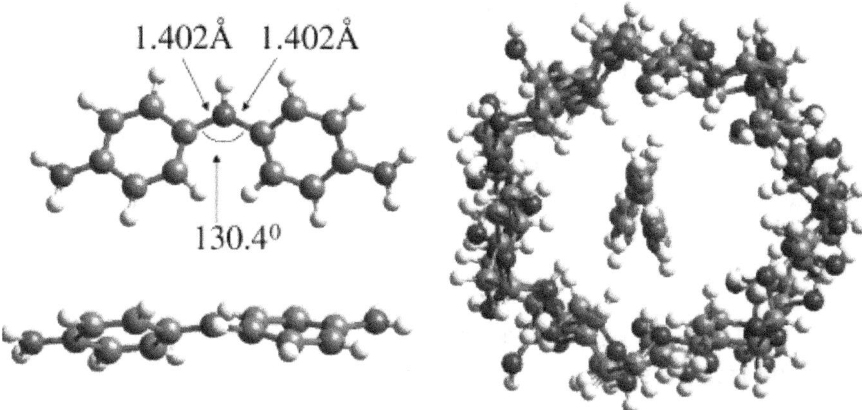

Figure 6.16   Structural analysis of polyaniline in β-cyclodextrin.[20]
Reproduced by permission of The Japan Society of Applied Physics.

own *N*-layered integrated molecular orbital and molecular mechanics (ONIOM) method, including both quantum mechanics and molecular mechanics calculations.[20] The calculation results indicated that the most stable structural configuration of polyaniline in β-cyclodextrin was near-planar geometry, of which the electronic configuration was almost the same as that in the planar conformation, as shown in Figure 6.16. And no charge transfer between polyaniline and β-cyclodextrin was observed, which suggests that β-cyclodextrins behave as insulating molecules. Consequently, polyaniline in the insulated molecular wire has extended conformation with highly conjugated electronic structure since it is confined in the small cavity of β-cyclodextrin, and thereby the isolated near planar configuration is stabilized.

The structural configuration of insulated molecular wires of polythiophene and α- or β-cyclodextrins was also calculated by the combined quantum mechanics/molecular mechanics (QM/MM) method.[21] Similarly, it turned out that the configuration of polythiophene in α- and β-cyclodextrins, or the cross-linked molecular tube was close to a planar structure. There was no charge transfer between the cyclodextrins and polythiophene, and the most stable electronic structure of polythiophene in cyclodextrins was near planar geometry, almost the same as that of the planar conformation of polythiophene in free space. This indicates that the polythiophene-based pseudopolyrotaxane can be regarded as insulated molecular wire, and the confinement of cyclodextrins effectively stabilizes the isolated near-planar configuration of polythiophene.

Next, the electronic transport through a single insulated molecular wire of polythiophene and β-cyclodextrin was investigated by the nonequilibrium Green's function formalism of quantum transport and density functional theory (DFT) of electronic structures with local orbital basis sets.[22] The calculation results indicated that the attachment of Au clusters on the insulated molecular wire did not modify the electronic structures of polythiophene confined in the cyclodextrin-based molecular tube, and hence did not affect the conductance properties of poly-thiophene. This means that the conduction path in the insulated molecular wire was protected from the outer environment by encapsulating cyclodextrins.

6.4 Optical Properties of Inclusion Complex Formation

Conjugated conducting polymers have been applied to optical devices for electroluminescence in recent years. Anderson and Frampton investi-gated the optical properties of insulated molecular wires of cyclodextrin-based polyrotaxanes with backbone conjugated conducting polymers.[4] Encapsulation with cyclodextrins can influence ultraviolet/visible (UV/vis) absorption and fluorescence spectra of the insulated molecular wires because the screening affects the outer environment around the backbone conjugated polymers, and the confinement stretches the backbone polymers to a rod-like conformation. In practice, spectral changes, such as sharper absorption spectra and sharper blue-shifted emission spectra than those of the uncovered conjugated polymers, were observed in the insu-lated molecular wire, as shown in Figure 6.17. These results indicates that encapsulation with cyclodextrins restricts reorganization in the excited state, including structural reorganization of the conjugated conducting polymer and/or reorganization of the solvent shell. Such optical change should strongly depend on how densely the conjugated conducting poly-mer is encapsulated with the rings.

It is widely known that insulated molecular wires enhance quantum yields for fluorescence compared to uncovered conducting polymers. This is often ascribable to encapsulating molecules restricting conformation of the backbone polymers and reducing flexibility of the excited state, similar to the blue-shifted emission. Encapsulation also prevents aggre-gation of conjugated conducting polymers, and energy or electron transfer between conducting polymers and environmental polar solvent or chro-mophores, which can quench fluorescence substantially. Anderson and Frampton found that cyclodextrin-based polyrotaxanes with backbone conjugated polymers, such as poly(paraphenylene), polyfluorene, poly(4,4′-diphenylenevinylene), and poly(phenylenevinylene), had few-fold higher fluorescence efficiencies than those of the backbone polymers both in solution and in the solid state.[4] This is mainly because encapsulation with cyclodextrins keeps the axis conducting polymers away from one

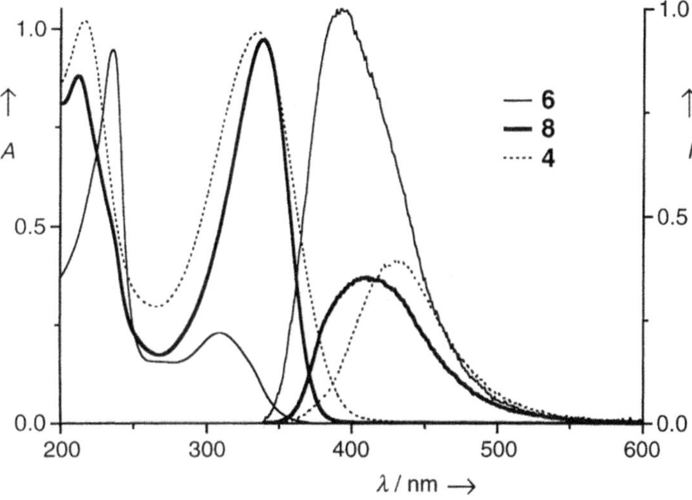

Figure 6.17   Absorption spectra (left) and emission spectra (right) of [2]rotaxane (6, solid), polyrotaxane (8, bold), and pseudopolyrotaxane (4, dashed) in aqueous solutions.[5c]
Reproduced by permission of John Wiley & Sons, Inc.

another and/or solvents, and thereby the insulated molecular wires hold the excited state over a longer term than the free polymers uncovered by cyclodextrins.

The high photoluminescence efficiency of insulated molecular wires can also be applied to organic light-emitting diodes (OLEDs).[4] Anderson and Frampton investigated electroluminescence in collaboration with Cacialli's group. The electroluminescence efficiency was enhanced substantially compared to uncovered conducting polymers in the cyclodextrin-based polyrotaxane with backbone conjugated polymers, such as poly(paraphenylene), polyfluorene, poly(4,4'-diphenylenevinylene) (PDV), and poly(phenylenevinylene), as shown in Figure 6.18. As a result, polyrotaxane OLEDs show higher activation voltage thresholds and brighter emission at the same voltage than those with the conjugated conducting polymers only. Blending the insulated molecular wires with poly(ethylene oxide) further increases the electroluminescence efficiency by a factor of up to 160.[23] This is ascribed to strong binding between the insulated molecular wires and poly(ethylene oxide), with complexation of the lithium counterions. Poly(ethylene oxide) wraps the insulated molecular wires to enhance the screening effect on the axis conjugated conducting polymers, and mobilizes the lithium cations to facilitate the charge transport and injection. These results indicate that the insulated molecular wires are promising materials for opto-electronic applications.

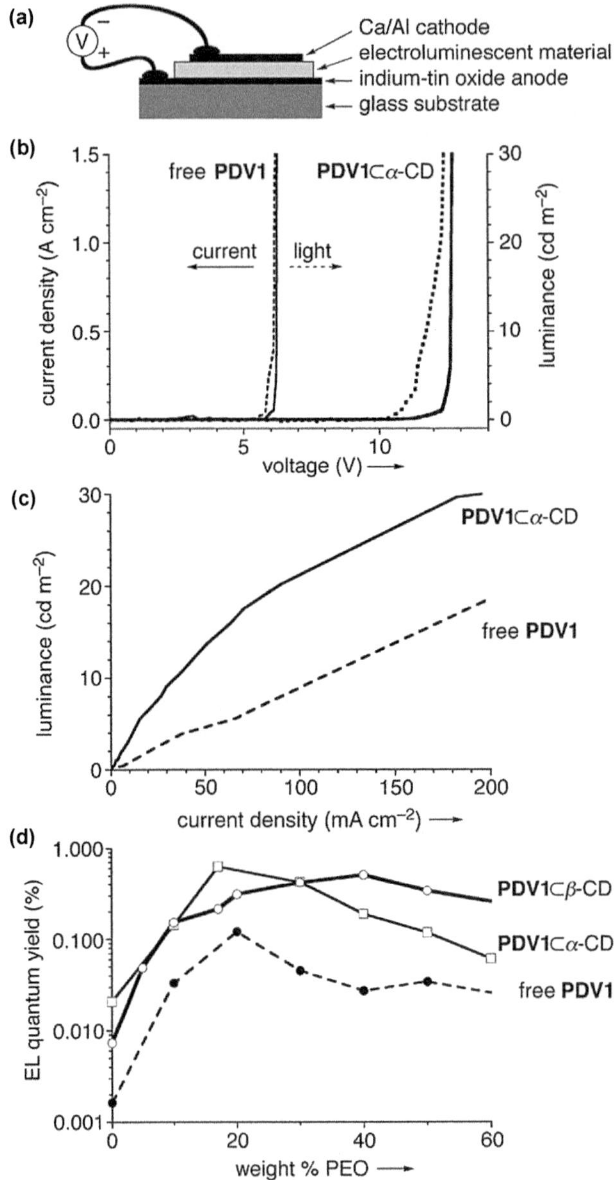

Figure 6.18 (a) Structure of a polyrotaxane OLED. (b) Variation in current density (—) and luminance (- - -) with voltage for typical OLEDs fabricated from PDV1 ⊂ α-CD and PDV1. (c) Data from (b) replotted as luminance *versus* current density. (d) Variation in electroluminescence efficiency with the weight fraction of poly(ethylene oxide) (PEO) for blends of PDV1 ⊂ α-CD, PDV1 ⊂ β-CD, and PDV1.[4]
Reproduced by permission of John Wiley & Sons, Inc.

References

1. H. Shirakawa, E. J. Louis, A. G. MacDiarmid, C. K. Chiang and A. J. Heeger, *J. Chem. Soc. Chem. Commun.*, 1977, 578; H. Shirakawa, *Angew. Chem.*, 2001, **113**, 2641; *Angew. Chem., Int. Ed.*, 2001, **40**, 2574.
2. J. H. Burroughes, D. D. C. Bradley, A. R. Brown, R. N. Marks, K. Mackay, R. H. Friend, P. L. Burn and A. B. Holmes, *Nature*, 1990, **347**, 539.
3. M. Maciejewski, *J. Macromol. Sci., Chem.*, 1979, **13**, 1175.
4. M. J. Frampton and H. L. Anderson, *Angew. Chem., Int. Ed.*, 2007, **46**, 1028.
5. (a) P. N. Taylor, M. J. O'Connell, L. A. McNeill, M. J. Hall, R. T. Aplin and H. L. Anderson, *Angew. Chem.*, 2000, **112**, 3598; (b) P. N. Taylor, M. J. O'Connell, L. A. McNeill, M. J. Hall, R. T. Aplin and H. L. Anderson, *Angew. Chem., Int. Ed.*, 2000, **39**, 3456; (c) J. J. Michels, M. J. O'Connell, P. N. Taylor, J. S. Wilson, F. Cacialli and H. L. Anderson, *Chem. – Eur. J.*, 2003, **9**, 6167; (d) J. Terao, A. Tang, J. J. Michels, A. Krivokapic and H. L. Anderson, *Chem. Commun.*, 2004, 56.
6. C. Lagrost, K. I. C. Ching, J.-C. Lacroix, S. Aeiyach, M. Jouini, P.-C. Lacaze and J. Tanguy, *J. Mater. Chem.*, 1999, **9**, 2351.
7. Y. Takashima, Y. Oizumi, K. Sakamoto, M. Miyauchi, S. Kamitori and A. Harada, *Macromolecules*, 2004, **37**, 3962.
8. I. Yamaguchi, K. Kashiwagi and T. Yamamoto, *Macromol. Rapid Commun.*, 2004, **25**, 1163.
9. M. van den Boogaard, G. Bonnet, P. van't Hof, Y. Wang, C. Brochon, P. van Hutten, A. Lapp and G. Hadziioannou, *Chem. Mater.*, 2004, **16**, 4383.
10. J. Terao, *Polym. Chem.*, 2011, **2**, 2444.
11. J. Terao, Y. Tanaka, S. Tsuda, N. Kambe, M. Taniguchi, T. Kawai, S. Saeki and S. Seki, *J. Am. Chem. Soc.*, 2009, **131**, 18046.
12. A. G. MacDiarmid in *Conjugated Polymers and Related Materials*, ed. W. R. Salaneck, I. Lundström, B. Rånby, Oxford University Press, Oxford, 1993.
13. K. Yoshida, T. Shimomura, K. Ito and R. Hayakawa, *Langmuir*, 1999, **15**, 910.
14. N. Ookubo, Y. Hirai, K. Ito and R. Hayakawa, *Macromolecules*, 1989, **22**, 1359.
15. T. Shimomura, H. Sato, H. Furusawa, Y. Kimura, H. Okumoto, K. Ito, R. Hayakawa and S. Hotta, *Phys. Rev. Lett.*, 1994, **72**, 2073.
16. T. Shimomura, K. Akai, T. Abe and K. Ito, *J. Chem. Phys.*, 2002, **116**, 1753.
17. A. Harada, J. Li and M. Kamachi, *Nature*, 1993, **364**, 516.
18. T. Akai, T. Abe, T. Shimomura and K. Ito, *Jpn. J. Appl. Phys.*, 2001, **40**, L1327.
19. T. Shimomura, T. Akai, M. Fujimori, S. Heike, T. Hashizume and K. Ito, *Synth. Met.*, 2005, **153**, 497.
20. R. V. Belosludov, H. Mizuseki, K. Ichinoseki and Y. Kawazoe, *Jpn. J. Appl. Phys.*, 2002, **41**, 2739.

21. R. V. Belosludov, H. Sato, A. A. Farajian, H. Mizuseki, K. Ichinoseki and Y. Kawazoe, *Jpn. J. Appl. Phys.*, 2003, **42**, 2492.
22. R. V. Belosludov, A. A. Farajian, H. Mizuseki, K. Ichinoseki and Y. Kawazoe, *Jpn. J. Appl. Phys.*, 2004, **43**, 2061.
23. J. S. Wilson, M. J. Frampton, J. J. Michels, L. Sardone, G. Marletta, R. H. Friend, P. Samorì, H. L. Anderson and F. Cacialli, *Adv. Mater.*, 2005, **17**, 2659.

Synthesis of Polyrotaxane and Slide-ring Materials

7.1 General Strategy for Synthesis

Characteristic properties and functions of polyrotaxanes and slide-ring materials can be designed through the synthesis of the polyrotaxane and subsequent cross-linking. Thus, the diversity of the chemical structure of polyrotaxanes is crucial for the material design. In this chapter, general strategies for the synthesis of polyrotaxane are introduced with perspective for the diversification and simplification.

For the synthesis of a polyrotaxane that consists of a single polymer chain and several cyclic components, there are three strategies, as shown in Scheme 7.1, so-called "clipping", "slipping", and "threading".

For clipping, both ends of the polymer are modified with bulky groups in advance, and then cyclization reactions of the pre-cyclization architecture are carried out on the polymer. Since cyclization should occur selectively on the backbone polymer, strong and specific interactions between the backbone and the pre-cyclization molecules are required. Because of the development of host–guest chemistry, diverse interactions are now available for the design of various interlocked supramolecular architectures beyond polyrotaxanes. Strong coordination bonds and strong hydrogen bonds between secondary ammonium and ethers are usually used (Figure 7.1).[1] The versatility is proved by the synthesis of a very complex interlocked architecture, so-called molecular Borromean rings.[2] In this way, this clipping method is based on the accumulated knowledge of host–guest chemistry over a long history. Therefore, this method excels in the syntheses of diverse polyrotaxanes that have precisely defined structures compared to the other two methods.

The slipping method is the most limited but the most convenient one. This method relies on the size agreement between the end groups of the host

Monographs in Supramolecular Chemistry No. 15
Polyrotaxane and Slide-Ring Materials
By Kohzo Ito, Kazuaki Kato and Koichi Mayumi

Published by the Royal Society of Chemistry, www.rsc.org

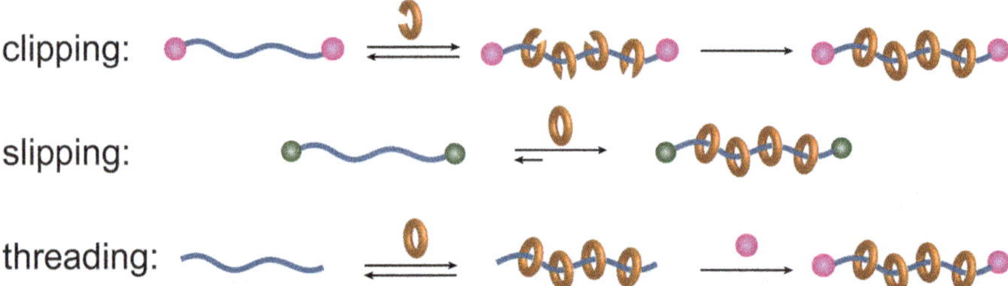

Scheme 7.1   The typical three strategies for the synthesis of polyrotaxane.

polymer and the cavity size of the cyclic components (Figure 7.2).[3] Because bulky end groups increase the energy barrier for the threading, the threaded cyclic components are not easily unthreaded once the threading has occurred. Heating of the system increases the chance to thread, and then cooling obtains the threaded structure. Although the interlocked structure can be dissociated by heating, the structure is stable at operating temperatures. The only difference is the way that thermal decomposition occurs: dethreading results in the decomposition of polyrotaxanes prepared by the slipping method, whereas bond cleavage initiates that of polyrotaxanes prepared by the other two methods.

Threading is the most realistic method for the cyclodextrin (CD)-based polyrotaxane. Although CDs have no specific binding cite, unlike crown ethers, the hydrophobic inner surface of the rings can take various polymers inside of the cavities. The driving force is generally regarded as one of the hydrophobic interactions, but the hydrogen-bonding networks formed by water molecules inside the cavity also contribute to the inclusion significantly.[4] At the same time, the hydrogen bonds between CDs threaded are also recognized to be the main factor that accelerates inclusion.[5] These two forces come from the nature of CDs, and thus the threading is observed in various polymers with size complementarity with regards to the CD cavities and the thickness of the guest polymers. Table 7.1 shows a list of observed inclusion for representative polymers depending on the sizes of the CDs.[6] Thin polymers without side chains, such as poly(ethylene glycol) (PEG) tend to form inclusion complexes with α-CD. On the other hand, thick polymers having side chains such as poly(methyl methacrylate) (PMMA) are thicker than the cavity size of α-CD, and thus they can form complexes only with γ-CD. Notably, thin polymers prefer α-CD rather than larger CDs, despite their sufficient cavity sizes for capturing such thin polymers. In addition, thin polymers can be included in the large cavity of γ-CD only when two polymer chains thread in a single cavity. This size-selective complexation clearly indicates that the match of the cavity size and the thickness of the polymer can stabilize the complexes, as known in the host–guest chemistry of CDs with small guests.[7] By selecting suitable CDs for each polymer, we can obtain inclusion complexes between CDs and polymers, so-called pseudopolyrotaxanes, just by mixing them together in water. Subsequent end-capping for both ends of the

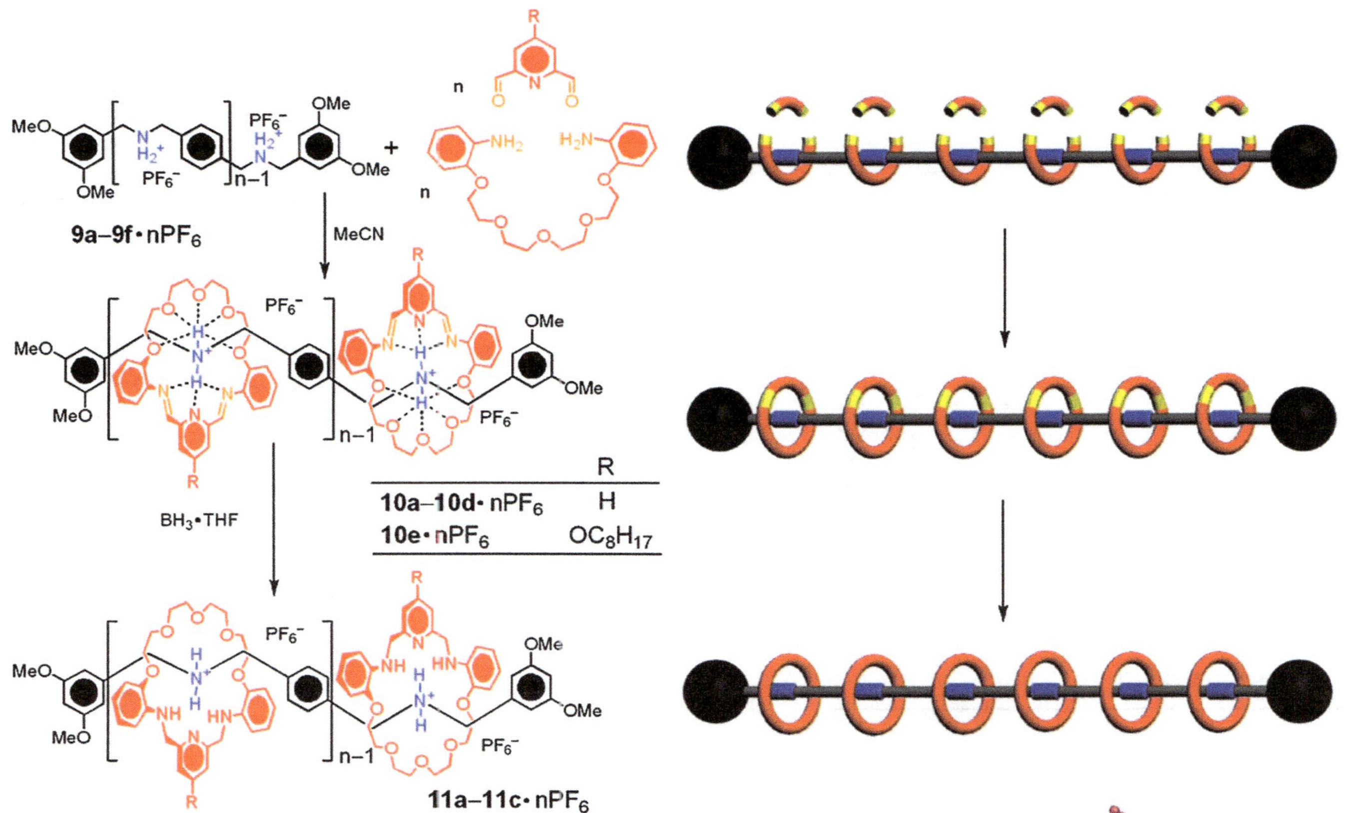

Figure 7.1 Example of the clipping method.
Reproduced from Ref. 1 by permission of The Royal Society of Chemistry.

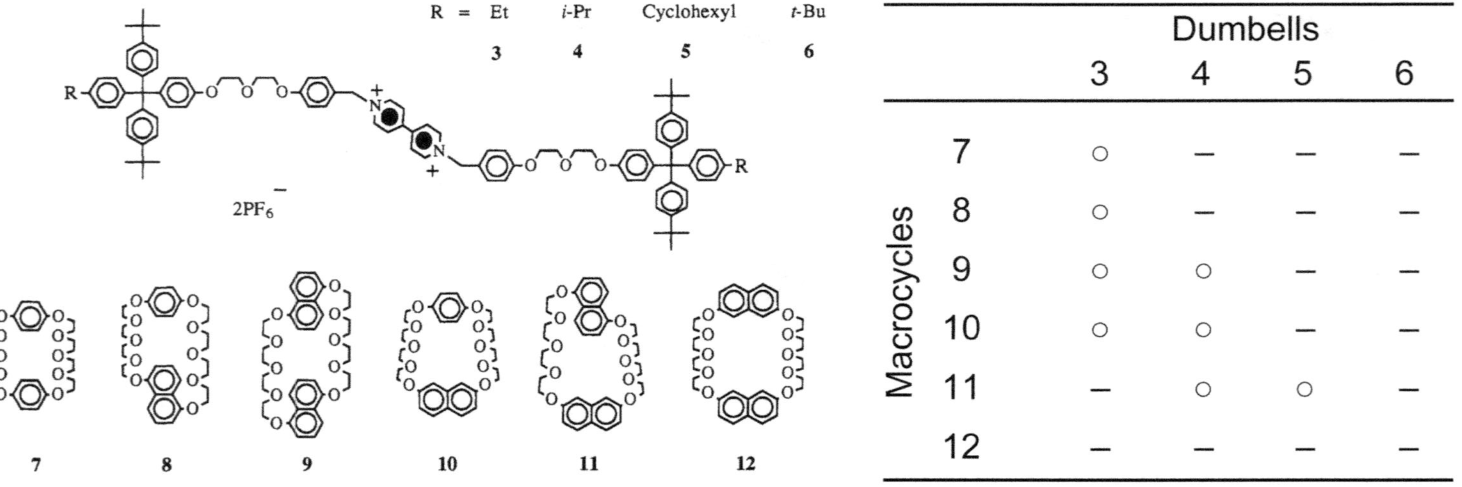

Macrocycles	Dumbells			
	3	4	5	6
7	○	−	−	−
8	○	−	−	−
9	○	○	−	−
10	○	○	−	−
11	−	○	○	−
12	−	−	−	−

Figure 7.2 Size-complemental formation of polyrotaxanes by slipping. The table indicates the formation of polyrotaxanes (○) or not (−).[3] Reproduced by permission of IUPAC.

Table 7.1 Inclusion complex formation ($\bigcirc$) or not (–) between various polymers and CDs.

Guest polymer	α-CD	β-CD	γ-CD	Guest polymers	α-CD	β-CD	γ-CD
Poly(ethylene glycol)	$\bigcirc$	–	$\bigcirc^a$	Poly(L-lactic acid)	$\bigcirc$	–	–
Polypropylene glycol	–	$\bigcirc$	$\bigcirc$	Polyethylene	$\bigcirc$	–	–
Polytetrahydrofuran	$\bigcirc$	–	$\bigcirc$	Polypropylene	–	$\bigcirc$	$\bigcirc$
Poly(methylvinyl ether)	–	–	$\bigcirc$	Polyisobutylene	–	$\bigcirc$	$\bigcirc$
Poly(ethylvinyl ether)	–	–	$\bigcirc$	Polybutadiene	–	$\bigcirc$	$\bigcirc$
Polyethyleneimine	$\bigcirc$	–	$\bigcirc$	Polybutadiene (1,4-)	$\bigcirc$	–	–
Poly(ethylene adipate)	$\bigcirc$	–	$\bigcirc$	Nylon 6	$\bigcirc$	$\bigcirc$	–
Poly(ε-caprolactone)	$\bigcirc$	–	$\bigcirc$	Silicone (polydimethylsiloxane, PDMS)	–	–	$\bigcirc$

[a]Double-stranded pseudopolyrotaxane is formed under special conditions.

threading polymers can produce polyrotaxanes, though the characteristic motions, such as sliding, are frozen without activation (see Section 7.2.4).

In this way, unless we are very skilled supramolecular chemists, it is likely that the threading method is the most useful for synthesizing polyrotaxanes. Since fine reviews for the clipping methods are available, and the method is not realistic for CD-based polyrotaxanes,[1,8] we show here the detail of the polyrotaxane synthesis by using the threading methods.

7.2 Synthesis of CD-based Polyrotaxane by the Threading Method

Although threading has big advantages in the formation of inclusion complexes, the subsequent end-capping reaction is necessary to obtain polyrotaxane. In addition, even after end-capping, another process is usually necessary to "activate" the polyrotaxane. The scheme in Figure 1.1 is too simplified to represent the elemental steps. Scheme 7.2 shows the general process to synthesize "active" polyrotaxane, consisting of (1) activation of the end groups of guest polymers, (2) inclusion complex formation with CDs, (3) end-capping of the complex, and (4) modification of the CDs to "activate" the characteristic intramolecular motions in polyrotaxanes such as sliding and rotation of CDs.

7.2.1 Activation of the End Groups of Guest Polymers

For activation of polymer ends, the end-capping process should be considered at the same time. As mentioned in Section 7.2.3, there are more serious demands for the end-capping process. Because there are diverse functional groups that can be introduced to the polymer ends, we should determine the groups at the polymer ends lastly. It is considered that the

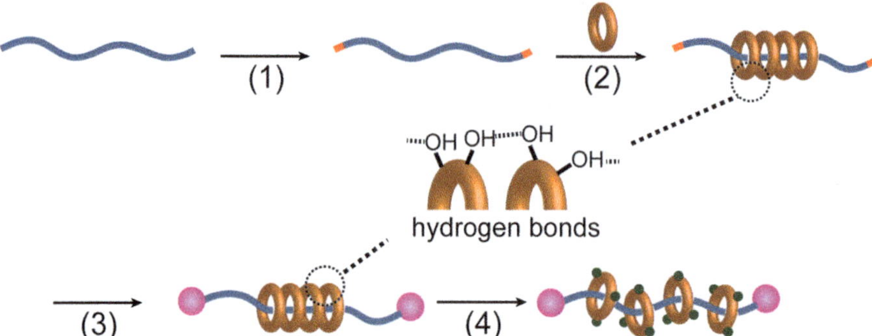

Scheme 7.2   General scheme for the threading method of polyrotaxane synthesis that consists of four elemental steps. (1) Activation of the end groups of guest polymers, (2) inclusion complex formation with CDs, (3) end-capping of the complex, and (4) modification of the CDs to "activate" the characteristic intramolecular motions.

functional groups should not disturb the inclusion complexation with CDs. Thus, simple and small functional groups, such as primary amines and carboxyl groups, are favorable and frequently employed. Some polymers, such as polyethylene, have difficulties in the functionalization of the end groups. For the synthesis of polyrotaxanes composed of such polymers, posterior conversions of the backbone polymers of polyrotaxanes are available (see Section 7.4.2.2).

7.2.2   Inclusion Complex Formation with CDs

The inclusion process is the key for this threading method. As mentioned in Section 7.1, the advantage is the spontaneous complex formation with CDs, particularly in water. Therefore, water-soluble polymers have an advantage due to the efficient and homogeneous process. In addition, homogeneous mixed solutions between CDs and guest polymers generally become turbid according to the progress of the inclusion. Although the hydrophobic interaction is the driving force for the first step, hydrogen bonds between threaded CDs are necessary for the propagation of the complex. A CD meets a polymer end to form an inclusion complex first, and then the CD should slide toward the center of the guest polymer to allow the next CD to be threaded at the polymer end. Unless the CD captured at the end of the polymer has enthalpic gain, the threading process is unfavorable energetically, though some triblock copolymers can progress the process with favorable middle blocks, *e.g.*, PEG–PPG–PEG for β-CD (PPG: polypropylene glycol). A theoretical calculation indicated that the hydrogen bonds largely contribute to the complexation.[4] As a result of the hydrogen bonds between threaded CDs, the CDs form a hexagonal crystal structure, as shown in Figure 7.3. Such crystal structures can be observed clearly by X-ray diffractometry, as shown in Section 5.1, thus the crystal structure is one of the

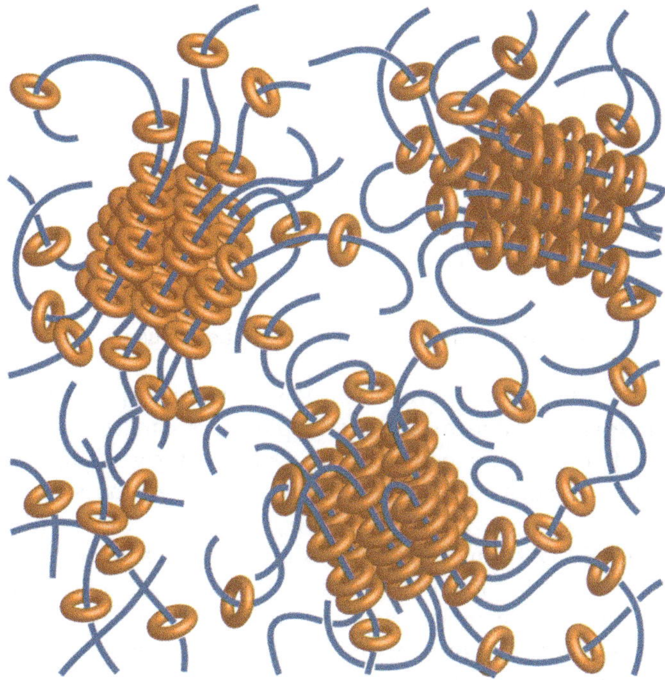

Figure 7.3 Crystal structure formed by CDs as a result of complexation.

simplest and most decisive indicators for the complexation. Because of the robust crystal structure formed by the threaded CD, the aqueous mixture of CDs and water-soluble polymers form the crystals, and the propagation of the crystals result in the precipitates. The precipitation of the complex can simplify the isolation of the complex from the free components; filtration and washing with water are usually enough. In the case of PEG and α-CD, almost quantitative complexation occurs. The efficient inclusion is the main advantage to achieve the mass production of polyrotaxane composed of PEG and α-CD.

However, the situation is not so simple for water-insoluble polymers. The complexation process is inevitably inhomogeneous. Such polymers are usually dissolved in organic solvents miscible in water, and the solution is mixed with an aqueous solution of CDs. Although precipitation is observed anyway, we can hardly distinguish the precipitation caused by desirable complex formation from that by reprecipitation of each component. Thus, the precipitation cannot be an indication for the complexation, unlike the system with water-soluble polymers. Such an inhomogeneous reaction naturally has a disadvantage with regards to efficiency, because the complexation occurs only at the surface of the aggregated polymers. Thus, efficient dispersion and stirring are needed for the complexation, though the reproducibility is obviously much lower than the homogeneous complexation. In addition, the isolation of the complex becomes much more difficult.

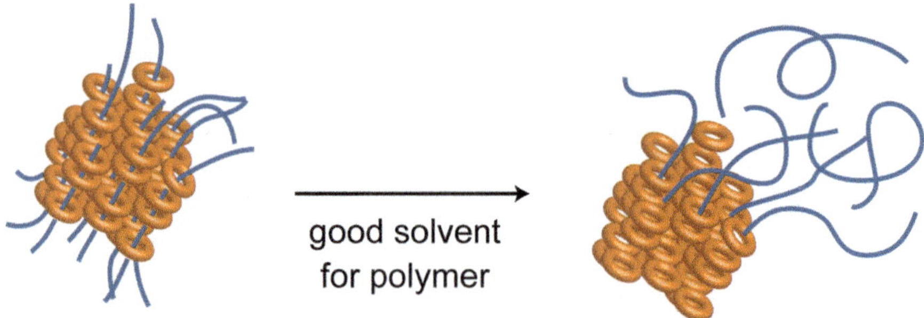

Figure 7.4   Extraction of backbone polymers from the inclusion complex.

Although the same filtration and washing with water is probably enough to eliminate free CDs, water-insoluble free polymers can remain in the precipitate. Good solvents for polymers are available for washing out the free polymer. However, such solvents often draw out the polymers threaded from the complex easily (Figure 7.4); for example, chloroform can extract all PDMS threaded with γ-CD from the precipitate of their crude complexes in 5 minutes. Therefore, the yield of the complex formation usually becomes low. To minimize this disadvantage, the crude inclusion complex can be used for the subsequent end-capping reaction without purification, though the resultant mixture should be freeze-dried to eliminate water that usually affects the capping reaction.

It is notable that this complexation process is the only process that relies on the spontaneous molecular recognition between host and guest, whereas the other three processes are a matter of synthetic organic chemistry. As mentioned in Section 7.1, there are many polymers that form no inclusion complex with CDs, despite the polymers being thin enough to be captured by CDs. This certainly shows that the complexation is a molecular recognition, and CDs reject some polymers not only for steric reasons. The limitation of the complex formation can be partly overcome by recent progresses in synthesis, and in mechanism elucidation. Some techniques for the synthesis of the inclusion complex and polyrotaxanes that have not been obtained simply by the above-mentioned spontaneous complexation are introduced in Section 7.4.2.2.

7.2.3   End-capping of the Inclusion Complex

End-capping is necessary to avoid the dissociation of the inclusion complex. The most common capping is achieved by introducing bulky groups at the polymer ends. This concept is based on the implication that threaded CDs cannot go through the polymer ends when the end groups are bulkier than the inner diameters of the CDs. However, how can we estimate the bulkiness of the end groups? It is not difficult for highly symmetric groups such as adamantane. On the other hand, planar groups found in aromatic groups typically make the situation more complicated. For example, let us think about the difference between *p*-nitrophenyl and *o*-nitrophenyl end groups.

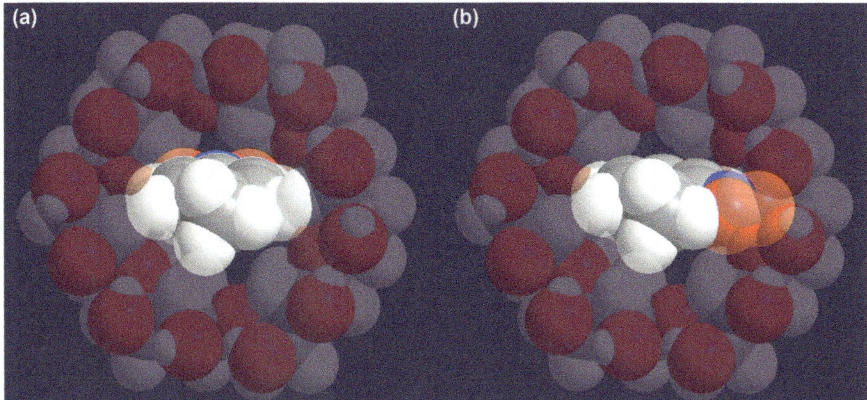

Figure 7.5 Difference in the bulkiness of capping groups for α-CD polyrotaxanes. (a) *Para-*, and (b) *ortho*-nitrophenyl groups. The nitro group at the *ortho* position toward the backbone makes the phenyl group bulkier, whereas that at the *para* position makes no difference.

Although the volumes of the both end groups are the same, the effect as the stopper is very different. From the viewpoint of the CD cavity, the nitro group at the *para* position made the phenyl group thicker, but it is unlikely to disturb the threading significantly (Figure 7.5(a)). On the other hand, the same substituent group at the *ortho* position widens the phenyl group, and the end group seems to disturb the threading obviously (Figure 7.5(b)). Thus, the circumcircle diameter can be a measure for the bulkiness of the end groups. When the diameter is larger than that of the cavity, the dissociation of the inclusion complex can be avoided.

However, such estimation can produce the minimum requisite, not the sufficient condition. Let us think again about the nitrophenyl end groups. From the estimation based on the circumcircle diameter, *ortho-* and *meta-*nitrophenyl groups should have the same effect as the capping group to avoid the dissociation of the complexes. In reality, the *meta*-nitrophenyl group has little function as a capping group; the *ortho*-nitrophenyl group completely avoids the threading of α-CD, whereas the *meta*-nitrophenyl group allows it to dissociate from the inclusion complex. A similar geometric isomerism effect also appeared in the distinctive threading kinetics of α-CD with methyl groups on pyridyl end groups at different positions.[9] This phenomenon also indicates that the kinetics depend on the direction of threading through the asymmetric cavity of CDs, which have a larger opening with secondary hydroxyl groups and a smaller opening with primary ones. In this way, it is hard to predict the necessary and sufficient bulkiness of end-capping groups. Therefore, the best way for selecting end-capping groups is probably learning from the experimental results in the literature. Scheme 7.3 shows a list of end-capping groups that have been verified for avoiding dissociation of polyrotaxanes with various sizes of CDs. To confirm the effectivity of a certain end group for capping, we can try a preliminary

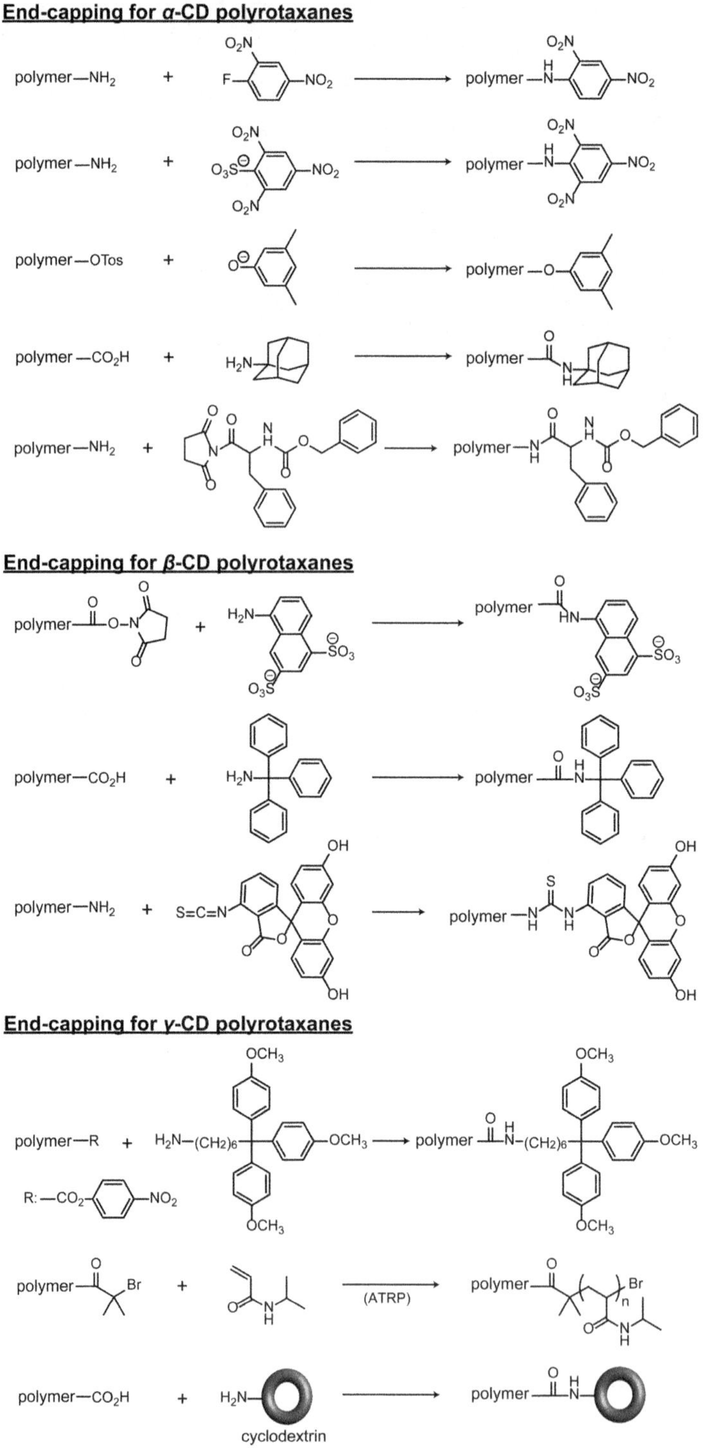

End-capping for *α*-CD polyrotaxanes

End-capping for *β*-CD polyrotaxanes

End-capping for *γ*-CD polyrotaxanes

Scheme 7.3 Examples of proven capping reactions for different sized CDs.

test by using an end-capped polymer. If complexation does not occur, it indicates that the end group is bulky enough for capping. Note that the complex formation sometimes arises from incomplete modification of the polymer ends. Anyway, this test is convenient for water-soluble polymers, because the complexation is visible as the precipitation mentioned previously.

The capping reaction tends to be more difficult with larger CDs, whereas γ-CD is the most versatile for inclusion of various polymers, as shown in Table 7.1. This is due to the less reactivity and less solubility of the end-capping reagents for larger CDs. For inclusion complexes with α-CD, phenyl groups with small substituents are sufficiently large. On the other hand, naphthalene groups are necessary for β-CD, and anthracene groups are even inadequate for γ-CD. Such larger aromatic groups tend to aggregate, showing poor solubility. In addition, larger reactants generally have less reactivity due to, for example, steric hindrance and less diffusivity. The nature of aggregation also makes the dissolution of the obtained polyrotaxane difficult, complicating the characterization of it. Indeed, the number of studies for polyrotaxanes with γ-CD is much less than those with α-CD. In addition, the synthesis of polyrotaxanes with γ-CD has another difficulty. As mentioned above, the unsubstituted anthracene is not bulky enough to avoid the dissociation of an inclusion complex with γ-CD. Even the larger triphenylmethyl (trityl) group is still inadequate. From molecular modeling, the circumcircle diameter of a trityl group is obviously larger than the inner diameter of γ-CD (Figure 7.6), though the *para*-substituted trityl group with, for example, a methoxy group becomes bulky enough to avoid dissociation. The

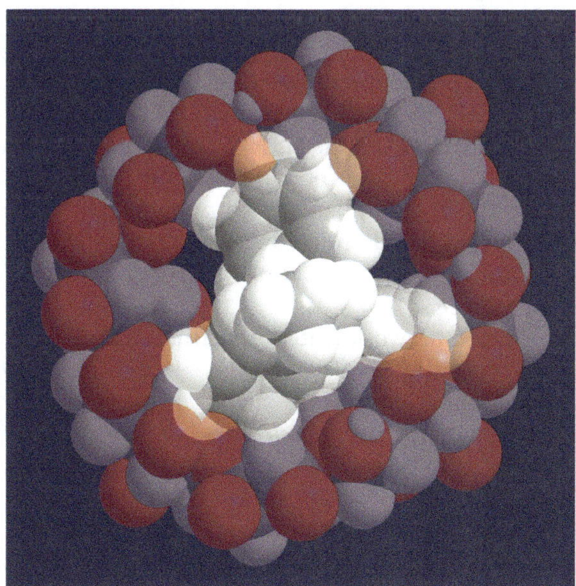

Figure 7.6 Comparison of the sizes of the cavity of γ-CD and a trityl group.

discrepancy probably arises from the static picture of the molecules. In reality, the shape of the CD cavity changes with the thermal fluctuation. The molecular structure of γ-CD is more flexible than smaller CDs, as the distorted rings can be observed in solution by nuclear magnetic resonance (NMR) and in the crystal structures. Thus, the reliability of the estimation for the bulkiness of the capping groups decreases with an increase of the cavity size. What is worse is that the number of successful examples is few. The confirmed end groups for γ-CD that are available for the simple threading method are substituted trityl groups and CD derivatives thus far. Recently, some techniques to overcome the difficulty in the capping reaction to obtain γ-CD polyrotaxanes were demonstrated, as mentioned in Section 7.4.2.1.

In addition to the bulkiness of the capping groups, the control of simultaneous dissociation of the inclusion complex is also necessary for the capping reaction. It is worthless if the reaction at the ends of the polymer is slower than the dissociation, producing only end-capped polymers like telechelic polymers without CDs threaded. One solution is using fast chemical reactions for the end-capping process. Actually, the first polyrotaxane was synthesized using one of the fastest chemical reactions between a primary amine and 2,4-nitro-fluorobenzene.[10] To accelerate the capping reactions, excess amounts of capping regents are usually used. However, there is no guideline for sufficiently fast reactions for the end-capping process, because the dissociation kinetics depend on both the host and guest molecules. It is notable that the dissociation kinetics also depend on the reaction conditions such as temperature and reaction solvents. Therefore, another solution is decelerating the dissociation of the complexes. As mentioned above, if the reaction solvents are good solvents for guest polymers, the polymers can be extracted from the cavity of the CDs. Thus, poor solvents for the polymers can prevent the dissociation. In the same way, poor solvents for the CDs can maintain their crystalline columnar packing. As long as the dissociation is significantly suppressed, the only requirement is the efficient chemical reactions of the capping reagent to the ends of the threaded polymers. For this strategy, finding poor solvents for both guest polymers and CDs is the first step. Then, we can line up candidates for capping reagents that are soluble in those solvents. Surely, the chemical reactions are limited by the solvents; for example, dehydration condensation reactions are not appropriate in water.

For a case study, let us work out the synthesis of a polyrotaxane that consists of polydimethylsiloxane (PDMS) and γ-CD (Figure 7.7). Non-polar solvents and very polar solvents are not appropriate, and intermediate solvents, such as methanol, acetone, and acetonitrile, are available. Because dehydration condensation between the carboxyl groups of the polymer ends and substituted trityl amines that are sufficiently reactive can be conducted efficiently even in an inhomogeneous system, alcohols are not appropriate. In this way, we can arrive at available chemical reactions and reaction solvents logically. Indeed, the condensation reaction between carboxyl-terminated PDMS and substituted trityl amines was conducted in acetonitrile

solvent	PDMS	γ–CD	capping regent
hexane	o	–	–
THF	o	–	o
acetonitrile	–	–	o
methanol	–	–	o
DMSO	–	o	o
water	–	o	–

Figure 7.7 Selection of appropriate solvent for the end-capping process of the PDMS/γ-CD complex. The table shows the solubility (o) or not (–) of the components and the capping regent.

to obtain the desired polyrotaxanes efficiently.[11] Incidentally, in the case of the most commonly used polyrotaxane that consists of PEG and α-CD, dimethylformamide (DMF) has been used traditionally from the first success of polyrotaxane synthesis. Although DMF is a good solvent for α-CD, the solvent cannot dissociate the inclusion complex because of the hydrogen bonds mentioned previously. Only in the presence of lithium salts, can the solvent dissociate it, as observed similarly in the dissolution of polysaccharides. DMF can be permeated into the complex to make it swollen, but not dissolve it. Such a swollen state has an advantage for the capping reaction compared to normal inhomogeneous reactions that occur only at the interface. The high efficiency of the capping reaction is one of the main reasons that the polyrotaxanes that consist of PEG and α-CD are the most commonly synthesized for study, and the only example of mass production. We should keep in mind that the reaction conditions are hardly applicable for other polyrotaxanes. A typical example is a polyrotaxane that consists of PEG and γ-CD. The inclusion complex is immediately dissolved in DMF. This is because the hydrogen-bonding network formed by the CDs is considerably weaker than the PEG/α-CD complex. The weakened hydrogen bonds appear in the intensity of the X-ray diffraction, and the improved solubility of the resultant polyrotaxane in DMF. Similar immediate dissociation occurs when the number of CDs threaded with a PEG backbone is considerably reduced, because of the weakened hydrogen bonds.

7.2.4 Activation of Intramolecular Motions

Although the product obtained after the end-capping process can surely be called a polyrotaxane structurally, the desired functions that originate from the intramolecular motions, such as sliding and rotation, are not observed. This kind of polyrotaxane should be called a "dormant" polyrotaxane. This next step is to release the motions to yield "active" polyrotaxanes and is necessary for crown ether polyrotaxanes as well (see the second step of Scheme 7.2); the strong binding between crown ethers and, for example, secondary ammonium groups on the guest polymers can be dissociated by the deprotonation of the ammonium. For CD-based polyrotaxanes, the hydrogen bonds between threaded CDs usually make the native poly-rotaxanes dormant. The simplest way to dissociate the hydrogen bonds is dissolving in good solvents. However, there are only a few good solvents for CD-based polyrotaxanes generally. As mentioned previously, the in-solubility arises from the strong hydrogen bonds between CDs. Dimethyl sulfoxide (DMSO) is the most commonly used solvents for polyrotaxanes composed of CDs and PEG. Since DMSO is a good hydrogen-bond acceptor, the solvent can interrupt the hydrogen bond (Figure 7.8). In addition, DMSO is a good solvent for both CDs and PEG. The good solubility for both components and the ability for interrupting hydrogen bonds should be satisfied. The mechanism of the insolubility of CD-based polyrotaxanes is very similar to that of polysaccharides, and the chemical structure of CDs as the major component of a polyrotaxane (in the case of PEG/α-CD poly-rotaxane, the CD is the major component with >70% wt%) resembles that of polysaccharides. Therefore, good solvents for polysaccharides that have been studied intensively can be utilized for the polyrotaxanes. Typical solvents for polysaccharides, including *N*-methylmorpholine *N*-oxide, lithium chloride/DMF (or dimethylacetamide, DMAc), an aqueous solution of Schweizer's reagent, and various ionic liquids, have been confirmed to dissolve polyrotaxanes.

Another method is the modification of hydroxyl groups of threaded CDs. This method is powerful, particularly for polyrotaxanes with hydrophobic backbone polymers. The above-mentioned solvents are effective to dissolve polyrotaxanes only when the solubility is governed by CDs. Even if such solvents can interrupt the robust hydrogen-bonding networks of CDs, the CDs are still immiscible to the hydrophobic backbones. For example, in the case of PDMS/γ-CD polyrotaxanes, there is no solvent that is good for both components (see imbedded table in Figure 7.7). Thus, the polyrotaxanes tend to form micelles. Fortunately, because CDs have modifiable functional

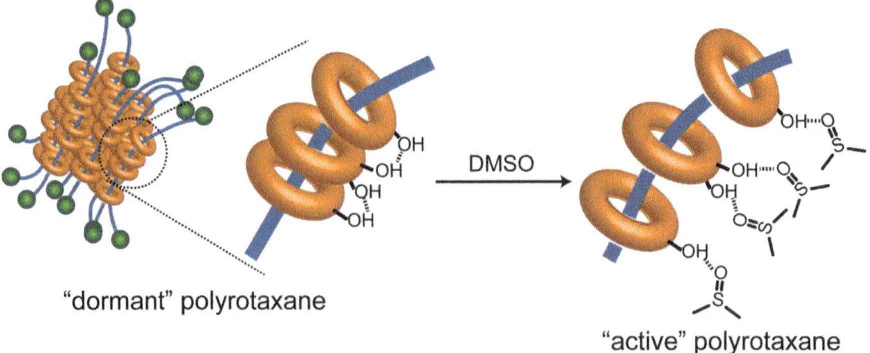

Figure 7.8   Activation of a PEG/α-CD polyrotaxane by intercepting the hydrogen bonds between threaded CDs with DMSO.

groups, the solubility of CDs can be tuned by the chemical modifications. However, the diversity of the modification of the CDs in polyrotaxanes is quite limited compared to that of free CDs. One reason for this is that the interlocked structure should be stored through the modification. Unlike free CDs, polyrotaxanes have functional groups on the backbone and on the end-capping groups. In particular, damage to the capping groups should be avoided, because the capping group is the keystone to prevent the dissociation of the interlocked structure. As mentioned previously, although we have various choices for the end-capping reactions, we should keep in mind that end-capping groups can withstand the subsequent chemical modifications to the cyclic components. For instance, the ester bond is convenient for the end-capping process, but modification of threaded CDs will be quite limited. In this way, if we start synthesizing polyrotaxanes without considering this activation process, we will face difficulties in the solubilization of polyrotaxanes, significantly complicating the characterization of polyrotaxanes.

7.3 Characterization of CD-based Polyrotaxane

The characterization of polyrotaxanes is completed by obtaining information about the following two aspects of polyrotaxanes. One is the characterization of the chemical components. A polyrotaxane is identified by 1) the chemical structure of the backbone polymer, 2) the chemical structure of the cyclic components, and 3) their stoichiometry, the so-called coverage or filling ratio. ^{1}H NMR spectroscopies are essential to determine the chemical structures and stoichiometry, similar to the characterization in conventional organic synthesis. Incidentally, the stoichiometry estimated from ^{1}H NMR is based on the assumption of single-stranded polyrotaxanes, and thus the characterization of double-stranded ones is much more complicated (Figure 7.9); the detail is mentioned in the last paragraph of this section. The second is the characterization of the interlocked structure. Even if two different components are detected by NMR, they are not always forming an inclusion complex, and can simply be a mixture. X-Ray diffractometry (XRD) is a powerful tool to distinguish the threaded structure from a simple mixture. As mentioned previously, the threaded CDs form columnar crystal structures, indicating a strong Bragg's peak of the (210) plane at $1.39\,\text{Å}^{-1}$

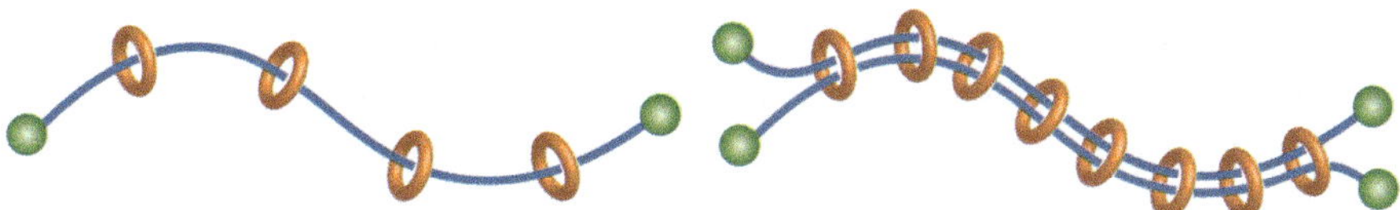

Figure 7.9 Single- and double-stranded polyrotaxanes that indicate the same stoichiometry by spectroscopy: the host–guest ratios are 4:1 and 8:2. Reproduced from Ref. 37 by permission of The Royal Society of Chemistry.

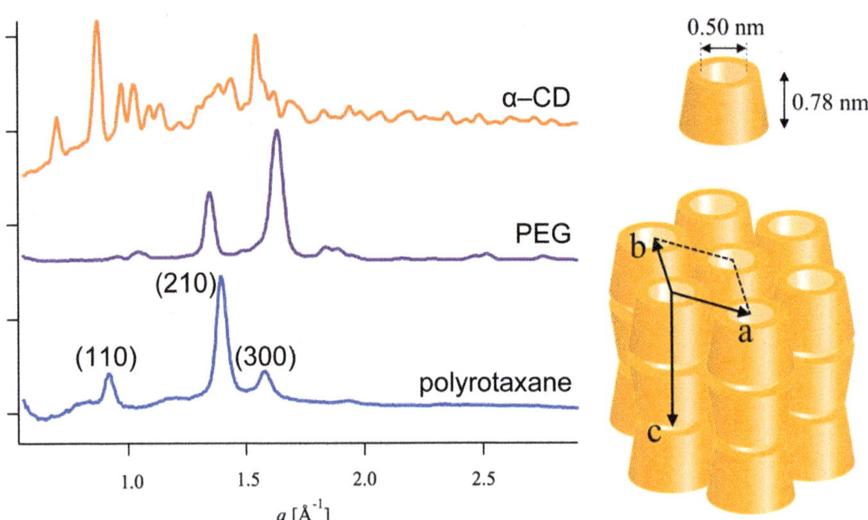

Figure 7.10   XRD patterns of a PEG/α-CD polyrotaxane and its components. The three characteristic peaks indicated were attributed to the columnar structure of threaded CDs.[12]
Reproduced by permission of IOP Publishing.

$(2\theta = 20°$ for CuKα) in the cases of α-CD and β-CD polyrotaxanes (Figure 7.10).[12] For γ-CD polyrotaxanes, a peak of the (420) plane at $1.13\,\text{Å}^{-1}$ $(2\theta = 16°)$ can be an indicator for the inclusion, though the signal is generally weaker than that of α-CD and β-CD polyrotaxanes. Although such crystal structures of the ring components can be indicators of the inclusion complexation, the crystal structures cannot distinguish polyrotaxanes from pseudopolyrotaxanes, which are just inclusion complexes without end-capping.

The easiest method to distinguish them is size-exclusion chromatography (SEC). Good solvents for both the polyrotaxane and each component should be used for the eluent. In diluted solutions (*ca.* 0.1%) in such good solvents, dissociation of the inclusion complex is more favorable. As a result, mere inclusion complexes or mixtures of polymers and CDs exhibit multiple peaks corresponding to the molecular weights of the individual components. On the other hand, polyrotaxane exhibits a single peak that corresponds to a molecular weight larger than each component. In this way, size-exclusion chromatography can distinguish polyrotaxanes from their components by the difference in molecular weight. However, this method is not available when there are no good solvents for polyrotaxanes and both components. The PDMS/γ-CD polyrotaxane is a good example. Since the polyrotaxane consists of hydrophilic CDs and a lipophilic polymer, there are no co-solvents, and the polyrotaxane is not soluble in simple solvents that are available for chromatography. In such a case, the modification of the cyclic components mentioned previously is sometimes useful for characterization.

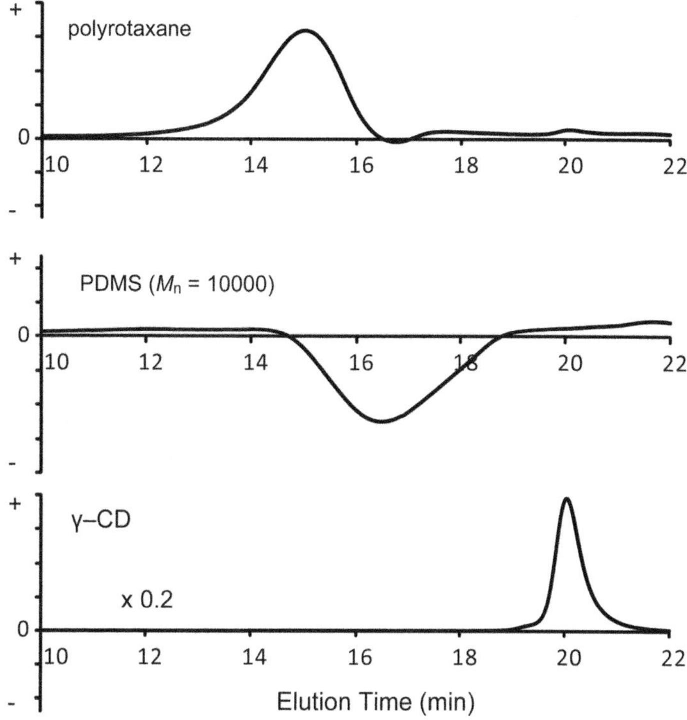

Figure 7.11 SEC chromatograms of an acetylated PDMS/γ-CD polyrotaxane and its components with chloroform as an eluent. Detection was by differential refractive index.[22]
Reprinted with permission from *Macromolecules*. Copyright (2009) American Chemical Society.

If lipophilic functional groups can be attached on the cyclic components without affecting the interlocked structure, the solubility of CDs can be converted to lipophilic enough to provide co-solvents for both the polyrotaxanes and each component. Acetylation of the CDs of PDMS/γ-CD polyrotaxane makes it soluble in non-polar solvents such as chloroform. The size-exclusion chromatogram using chloroform exhibits a peak that corresponds to a molecular weight higher than each component, which can also be an indication of impurities (Figure 7.11).

Mass spectrometry can be used for relatively small polyrotaxanes up to *ca.* 10–20 kDa. For insoluble and non-modifiable polyrotaxanes, this might be the only way to confirm the interlocked structure. If the polyrotaxanes are ionized without dissociation, the exact molecular weight can be determined unambiguously (Figure 7.12).[13] The mass spectrum can even determine the distribution of the number of CDs in the polyrotaxane. However, the optimization of the ionization is not very easy, and we might not be able to distinguish polyrotaxanes from mere pseudopolyrotaxanes. Solution NMR spectroscopies detecting the nuclear Overhauser effect (NOE) are also

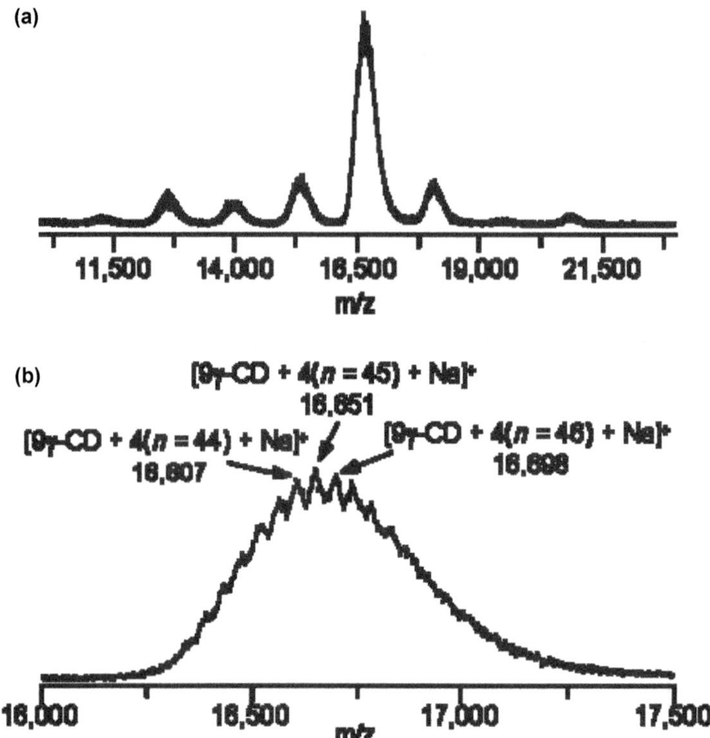

Figure 7.12   Matrix-assisted laser desorption/ionization-time of flight (MALDI-TOF) mass spectrum of a PEG/γ-CD polyrotaxane.[13] Constant cycles of peaks in every (a) 1297 Da, and (b) 44 Da that correspond to the molecular weights of γ-CD and the repeating unit of PEG are observed respectively.
Reprinted with permission from Macromolecules. Copyright (2009) American Chemical Society.

available sometimes for the confirmation of the interlocked structure. NOE can be observed when different protons are located closely; thus, this has been a powerful tool to detect the interactions between host and guest molecules. In the case of CDs, the protons at C_3 and C_5 are located inside of the inclusion cavities. Thus, the NOE between the protons of backbone polymers and the protons of CDs located in the inner cavity can be the indicator to conclude the polymers are threaded in the cavity of CDs (Figure 7.13).[14] Note that the NOE is not observed when the interaction is not sufficiently strong, and thus it is not always that no NOE means no inclusion complexation. In addition, we should be careful that the target proton is certainly excited selectively. When the chemical shifts of the CDs and guest polymer are overlapped, the selective excitation is very difficult. For example, when the proton of C_4 of CDs is overlapped with that of the target polymer, both protons are excited unselectively. As a result, correlation spectroscopy (COSY) signals from the protons of C_3 and C_5 of CDs are naturally observed. Such COSY signals appear like the desired NOE.

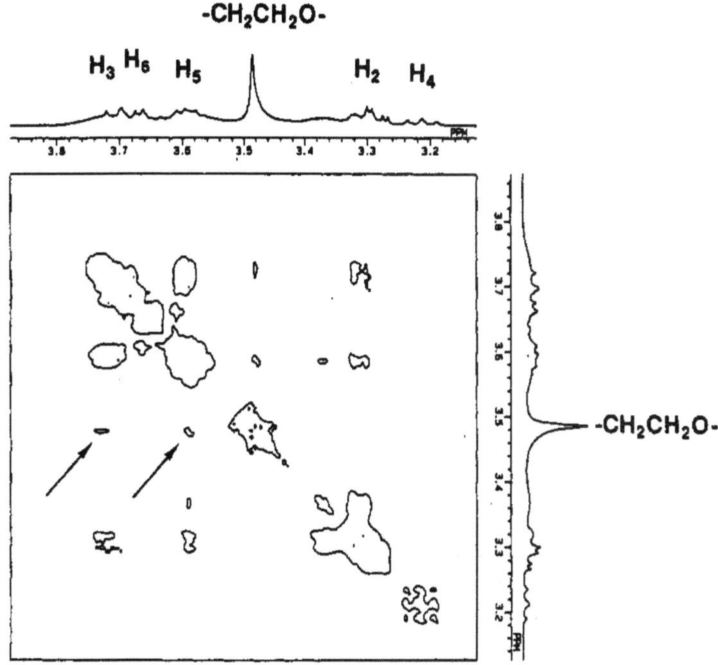

Figure 7.13 2-D NOE spectroscopy (NOESY) NMR spectrum of a PEG/α-CD polyrotaxane. Arrowed peaks indicated by arrows are the correlation peaks between a CD's protons inside of the cavity and those of the guest polymer.[14]
Reprinted with permission from *The Journal of Organic Chemistry*. Copyright (2009) American Chemical Society.

Determination of the stoichiometry is sometimes very complicated when single- and double-stranded inclusion complexes are possible. The first report of a double-stranded inclusion complex was the case between PEG and γ-CD.[15] The size of the cavity in γ-CD is large enough to be threaded with two PEG chains. It is notable that such double-stranded complexes are not always formed even if the thickness of the guest polymer is thin enough. For instance, both single- and double-stranded polyrotaxanes have been synthesized from PEG and γ-CD (see Section 7.4.3 for the detail). In such a case, ^{1}H NMR cannot discriminate between single- and double-stranded ones. Recall that both the densely packed double-stranded polyrotaxane (coverage = 100%) and half-covered single-stranded polyrotaxane (coverage = 50%) give the same molar ratio between the CD and the repeating unit of the backbone polymer (Figure 7.9). The finding of the double-stranded complex was supported by a systematic study comparing the complexation with α-CD and γ-CD. The α-CD/PEG polyrotaxane gave a ratio of 1 : 2, indicating a densely packed polyrotaxane, because the cavity of α-CD is not large enough to be threaded with two chains of PEG. On the other hand, the γ-CD/PEG gave a ratio of 1 : 4. This indicates either the half-covered single-stranded or

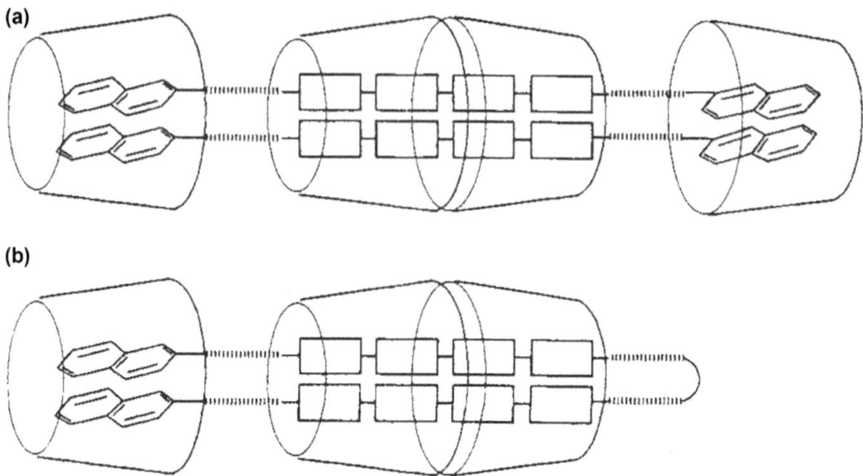

Figure 7.14 Two possible structures (a and b) of the double-stranded complex between α-CD and naphthyl-terminated PEG. π-Interactions between the end groups of PEG enable the double-stranded structure.[15]
Reprinted by permission from Macmillan Publishers Ltd: *Nature*, **370**, 126–128 copyright (1994).

the dense double-stranded complex. It is known that such dense complexes are generally formed with relatively short PEG (M_W < 2000) because of the efficient hydrogen bonds between the threaded α-CDs. Thus, it may be unnatural to replace γ-CD with α-CD and obtain exactly half coverage. In addition, the interaction between the end groups of PEG is observed in the diluted solution, indicating that two PEG chains are located closely, probably because of the double-threaded structure (Figure 7.14). The interaction was detected as the excimer's emission of the end groups, and was not observed in the complex with α-CD. In this way, the structure of the first double-stranded complex was determined. For more direct determination, the molecular weight of polyrotaxane would be decisive. However, as mentioned previously, reliable determination of the molecular weight is very limited. Mass spectrometry can be applied for relatively small polyrotaxanes. SEC can provide only the apparent molecular weight based on the calibration curves formed by standard polymers, not standard polyrotaxanes. Therefore, we should usually determine the stoichiometry of γ-CD-based polyrotaxanes from comprehensive data indirectly. See Section 7.4.3 for more details of single- and double-stranded polyrotaxanes that consist of PEG and γ-CD.

7.4 Diversification of the Chemical Structure of Polyrotaxane

Polyrotaxane has a diversity of chemical structure in 1) the backbone polymer, 2) the cyclic components, and 3) their stoichiometry, as mentioned in Section 7.3. However, there are many difficulties in the synthesis of polyrotaxane that prevent the diversification. In this section, some of synthetic strategies are introduced for the diversification of each component separately. Note that the process of each diversification started from the first polyrotaxane that consist of PEG and α-CD.

7.4.1 Diversity of the Cyclic Component

The diversification of the cyclic component started from modification of the hydroxyl groups of the CDs. Examples of the modified polyrotaxanes and their solubility are listed in Table 7.2.[16] As mentioned previously, the main advantages of CD-based polyrotaxanes are the versatile and efficient inclusion complexation with various polymers mediated by the hydrogen bonds between the hydroxyl groups of CDs. At the same time, modifications of the hydroxyl groups are usually required for dissolving polyrotaxanes. For PEG/α-CD polyrotaxane, hydroxylpropylation is the most commonly used method to solubilize the polyrotaxane in aqueous solutions. The strong orientation of the hydroxyl groups in the native CDs are weakened by the insertion of a propylene glycol unit as a linker. The advantage is that the number of the hydroxyl groups remains constant. As a result of the weakening of the hydrogen bonds, most of the co-solvents for PEG and α-CD, such as water and DMF, can dissolve the hydroxypropylated polyrotaxane as well. On the other hand, modifications of hydrophobic groups make the polyrotaxane soluble in hydrophobic solvents. Methylation, acetylation, and trimethylsilylation are often used to solubilize polyrotaxanes in, for example, tetrahydrofuran (THF), toluene, and chloroform. As long as the strong hydrogen bonds of the CDs are sufficiently weak, the solubility can be predicted from

Table 7.2 PEG/α-CD polyrotaxane derivatives with different functional groups on the cyclic component and their solubility.[16] Reproduced by permission of John Wiley & Sons, Inc.

Modification	Functional group	Reaction conditions[a]
Methylation	CH_3	Methyl iodide and t-BuOK in DMSO at r.t. overnight
Hydroxypropilation	$[CH_2CH(CH_3)_2]_nH$	Propylene oxide in 1 M aqueous NaOH at r.t. overnight
Tritylation	$C(C_6H_5)_3$	Trityl chloride and pyridine in DMSO at 80 °C overnight
Acetylation	$COCH_3$	Acetic anhydride with pyridine and DMAP in DMAc/LiCl at r.t. overnight
Trimethylsilylation	$Si(CH_3)_3$	HMDS in DMSO/THF at 60 °C for 1 day
Phenylcarbamation	$CONH–C_6H_5$	Phenylisocyanate with $Sn(Oct)_2$ in dry DMSO at r.t. overnight
Dansylation	$SO_2–Np(5\text{-}NMe_2)$	Dansyl chloride with triethylamine in DMAc/LiCl at 10 °C for 1 day
Nitration	NO_2	HNO_3/P_2O_5 at 0 °C for 20 min or BF_4NO_2 in CH_3CN at 0 °C overnight

[a]r.t., room temperature; DMAP, 4-dimethylaminopyridine; HMDS, hexamethyldisilazane; THF, tetrahydrofuran.

that of the corresponding free CDs found in the literature. The solubilization of polyrotaxane also makes it easy to characterize the chemical structure by using NMR and size-exclusion chromatography.

The diversification of the cyclic component has also been accelerated by strong demands for the utilization of the sliding ability of polyrotaxanes. The common strategy is to attach the functional molecules on the CDs of the polyrotaxane to increase the flexibility of the functional groups. For example, a polyrotaxane with CDs where selective-binding substrates are attached exhibited more efficient binding to enzymes immobilized on a substrate compared to the polymer where the substrates were attached directly (Figure 7.15).[17] Because the distance between substrates can be adjusted to the enzymes by the sliding, effective multivalent bindings are realized. A quick response of a liquid crystal polymer to an electric field was achieved by attaching a mesogen to the CDs of the polyrotaxane, rather than directly to the backbone polymer. In this way, the sliding ability has potential to amplify various functions.

Polymerizations of monomers from the cyclic components of poly-rotaxanes yield copolymers like graft-copolymers (Figure 7.16).[18] Unlike

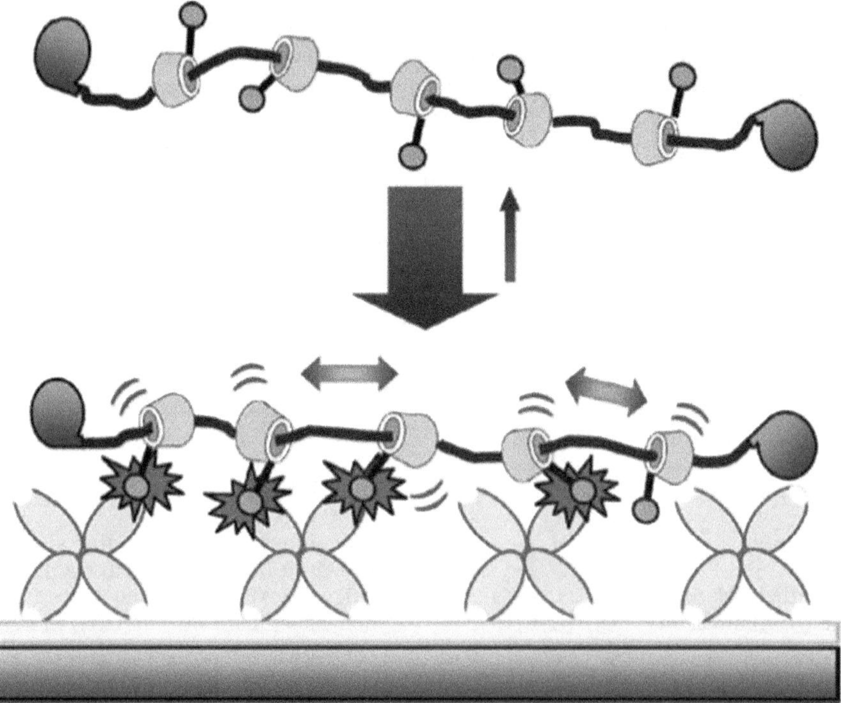

Figure 7.15   Efficient binding between substrates that are attached on the rings of a polyrotaxane and enzymes immobilized on the solid surface due to the sliding of rings.[17]
Reproduced by permission of John Wiley & Sons, Inc.

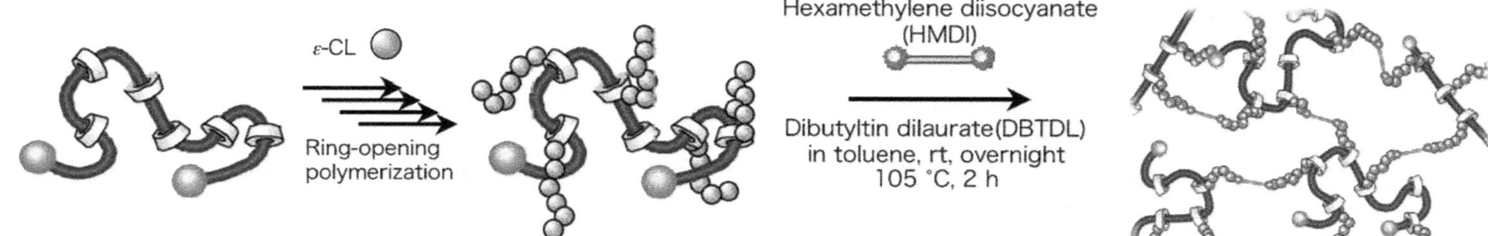

Figure 7.16 Synthetic scheme of a poly(ε-caprolactone)-grafted PEG/α-CD polyrotaxane and subsequent cross-linking to obtain a slide-ring elastomer.
Reproduced from Ref. 18 by permission of The Royal Society of Chemistry.

conventional graft-copolymers, the graft chains are not bound to the main chains but are topologically constrained in the molecules. These kinds of copolymers are called sliding graft-copolymers or graft-polyrotaxanes. For the grafting polymerization of side chains, reactions under mild conditions, such as ring-opening polymerizations and radical polymerizations, are often employed. Ring-opening polymerization of ε-caprolactone (ε-CL) can be carried out from the hydroxyl groups of CDs of polyrotaxanes. Radical polymerizations of vinyl monomers are also available. Notably, the "activation" of the polyrotaxanes by modifying the CDs is often necessary in advance for such grafting otherwise insoluble solids that cannot be characterized well are usually obtained. This is probably due to the uncontrolled polymerizations. Living radical polymerizations, such as atom transfer radical polymerization (ATRP), are promising methods to obtain various sliding graft-polymers.[19] For such polymerizations, initiators are necessary to be attached to the CDs of polyrotaxanes, and these reactions usually can be the "activation" of polyrotaxanes to improve solubility. The versatility of such living radical polymerizations enables us to design diverse sliding graft-polymers with various graft chains. The synthesis of such graft-polyrotaxanes is currently important industrially, as introduced in Section 8.4.

Contrary to attaching relatively small molecules to polyrotaxanes, there is another kind of molecular design that attaches polyrotaxanes to large matrixes. Slide-ring gels can be regarded as one of these designs. The coupling reaction of CDs from different polyrotaxanes is essentially the same as attaching polyrotaxanes on a matrix network made by polyrotaxanes. As a result of accumulation resulting from the attaching process, whole systems become infinite polymer networks, such as gels or elastomers. Similar to this concept, polyrotaxanes can be attached to other matrix networks instead of polyrotaxane networks. Copolymerization of vinyl monomers and polyrotaxanes that have CDs modified with vinyl groups can make polymer networks cross-linked by the polyrotaxanes (Figure 8.13). In this case, polyrotaxanes work as cross-linkers for the matrix polymers. Because the ability of chain sliding through the cross-links is retained in these systems, small amounts of polyrotaxanes can improve the mechanical strength of the matrix networks as described in Section 8.3. In addition, polyrotaxanes can be attached on substrates.[20] Although CDs of polyrotaxanes are immobilized on the surfaces of the substrates, the main chains are just topologically bound, and movable through the cavities of the immobilized CDs. This is an example of surface reforming, using the high mobility of the polymer chains of polyrotaxanes. Polyrotaxanes can be attached also on inorganic particles to obtain a new class of nano-composite material.[21] For instance, the sol-gel reaction of silyl-modified polyrotaxanes on the surface of inorganic particles can yield a nano-composite network. Unlike conventional nano-composites, the hardening of materials that are attributed to the reduction of the mobility of polymers can be suppressed.

In addition to the above-mentioned chemical modifications of the hydroxyl groups of the CDs of polyrotaxanes, CDs have a variety of ring sizes.

However, changing the CD size on polyrotaxanes is much more complicated. Because it seems to be almost impossible to change the size of the CDs of a certain synthesized polyrotaxane, the size of CDs should be determined prior to the synthesis. The important point is that we cannot determine the size of the CDs, but the backbone polymer determines their size spontaneously. As mentioned previously, the complexation process almost relies on molecular recognitions between guest polymers and host CDs. It is true that we can design the host and guest molecules to form their complexes by supramolecular chemical ways. However, the advantage of the CD-based polyrotaxane is in the simple and versatile complexations. Therefore, simple and versatile techniques, without changing the chemical structures of the CDs and polymers substantially, are required for making them form complexes spontaneously.

7.4.2 Diversity of the Backbone Polymers

As described in Section 7.4.1, the PEG/α-CD polyrotaxane has had a great impact on the applications of polyrotaxanes, advancing the diversification of the cyclic components. On the other hand, the backbone of the polyrotaxane is currently quite limited. The diversity of the backbones is important not only from the point of view of the applications but also from that of the intramolecular dynamics. We can imagine, for example, that the ability of chain sliding through CDs should depend on the chemical structures of the polymers. The main reason why the diversification has been delayed is due to the difficulty in polyrotaxane synthesis. Recall the four steps to obtain active polyrotaxanes for the threading method. The activation of polymer end groups has been studied well for various polymers, and the versatility of the inclusion complexation with CDs is supported by the diverse guest polymers. The problem is in the subsequent end-capping process.

7.4.2.1 Bulky Guest Polymers that Require γ-CD

According to the bulkiness (thickness) of guest polymers, the sizes of CDs should be adjusted. For PEG, α-CD is suitable, whereas β-CD or larger CDs are needed for polypropylene glycol (PPG). As a result of the increased sizes of CDs, bulkier end-capping groups are needed to prevent the dissociation of the interlocked structures. Such bulky capping reagents generally have problems with regards to their solubility, and thus good solvents are limited. As described in Section 7.2.3, the selection of reaction solvents is crucial for the end-capping process to prevent dissociation of the complex during the reaction. Thus, the available condition for end-capping is significantly limited. The situation is worse for polyrotaxanes with γ-CD. Unfortunately, as shown in Table 7.1, γ-CD is the most versatile from the aspect of complex formation. For many useful polymers, such as silicones, polydienes, and methacrylate polymers, γ-CD is the only potential host molecule. Therefore, reliable and versatile methods for the end-capping of inclusion complexes between such polymers and γ-CDs are desired.

Actually, examples of polyrotaxanes with γ-CDs synthesized by the threading method are significantly limited thus far. There are a few capping groups that have been confirmed bulky enough for polyrotaxanes with γ-CDs such as silsesquioxanes, *p*-substituted trityl derivatives, dimerized anthracene, and CDs. Theoretically, these groups can be bound to the ends of backbone polymers of the inclusion complexes to obtain polyrotaxanes. However, simple end-capping with bulky substituents is not generally valid for the synthesis. Only the *p*-substituted trityl derivatives were valid for the simple end-capping to yield PDMS/γ-CD polyrotaxanes, though the condensation reaction between carboxyl ends of PDMS and *p*-substituted tritylamine derivatives were not mediated by condensation reagents, and thus an active ester was used as the polymer ends.[22] Instead of simple capping, dimerization of anthracenyl groups was used (Figure 7.17).[23] Anthracenyl

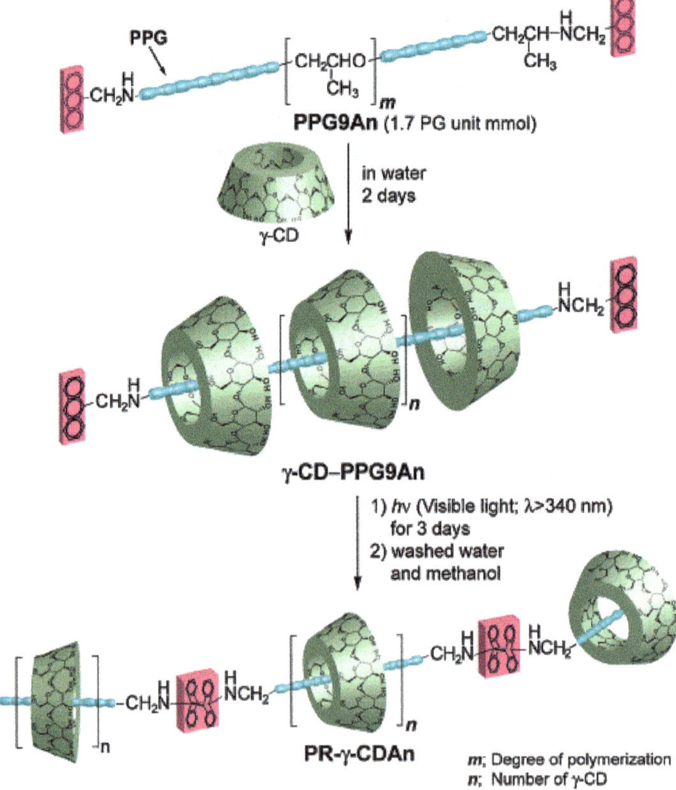

Figure 7.17   End-capping of the PPG/γ-CD inclusion complex by intermolecular photo-dimerization of the end groups of the backbone polymer.[23]
Reprinted with permission from *Macromolecules*. Copyright (2004) American Chemical Society.

groups are introduced to both ends of the backbone polymers prior to the inclusion complexation with γ-CD. The obtained inclusion complex was irradiated by visible light to dimerize the anthracenes at the ends of the backbones. This method is a kind of polymerization of pseudopolyrotaxanes. As a result, bulky joints are formed in the elongated backbone. γ-CDs between the joints are trapped to form polyrotaxanes. The advantage of this method is that capping reagents are not necessary, and thus the same conditions as the inclusion process are available to minimize the possibility of dissociation induced by, for example, changing reaction solvents. Because polymerizations are used for end-capping, the resultant polyrotaxanes exhibit a very broad distribution of molecular weight.

Polymerizations of bulky monomers initiated from the polymer ends of inclusion complexes are also available to obtain γ-CD polyrotaxanes.[24] For example, poly(*N*-isopropylacrylamide) can be grafted from the ends of backbone polymers of the inclusion complexes (Scheme 7.4). At first,

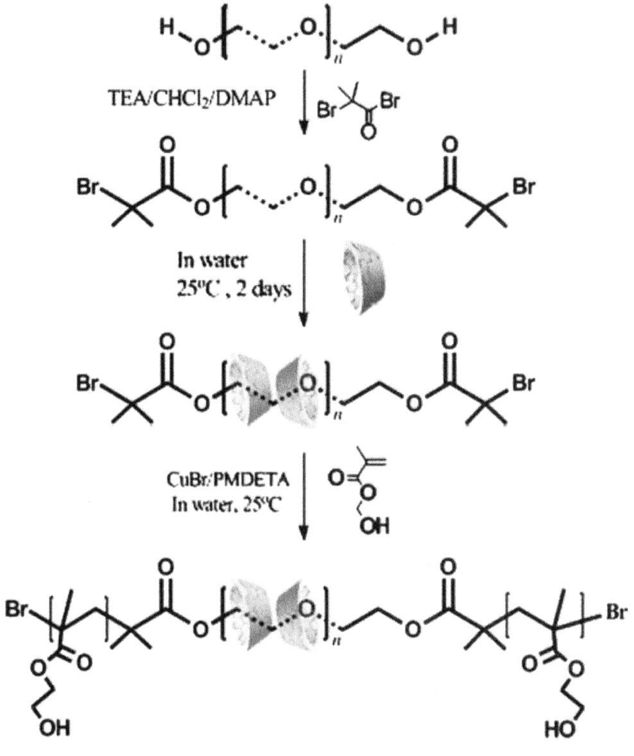

Scheme 7.4 An example of polyrotaxane synthesis *via* end-capping by ATRP of hydroxyethyl methacrylate (HEMA) from both ends of the backbone polymer (TEA, triethylamine; DMAP, 4-dimethylaminopyridine; PMDETA, pentamethyldiethylenetriamine).[24]
Reprinted from *Polymer*, **49**, 4489–4493, Copyright (2008), with permission from Elsevier.

macroinitiators are prepared from candidate polymers by chemical modification of their end groups with, for example, an ATPR initiator. Hydroxyl groups of the PEG–PPG–PEG tribrock copolymer are reacted with 2-bromoisobutyrate. Then, the activated polymer forms an inclusion complex with γ-CD by simply mixing each aqueous solution. Finally, the inclusion complex generated as a precipitate is end capped by the polymerization of *N*-isopropylacrylamide from the ends of the backbone polymers. As a result, the backbone polymer becomes a penta-block copolymer, and both end blocks of poly(*N*-isopropylacrylamide) (PNIPAAm) work as stoppers to prevent dissociation of the complex. Although the bulkiness of the monomer used for this method seems to be considerably less bulky than other capping groups, such as *p*-substituted trityl derivatives and CDs, the length of the prolonged polymers from the ends are significantly longer. This implies that not only the bulkiness but also the length of the end groups are important for preventing the dissociation of the inclusion complexes. It is notable that the obtained polyrotaxane has relatively large end groups that may affect various properties. Indeed, such large end groups considerably contribute to micelle formation.[25]

As long as the backbone polymer is thin enough to form an inclusion complex with β-CD or α-CD, the synthesis of polyrotaxanes that contain γ-CD is not very difficult. The key for this strategy is mixing small amounts of α-CD during the complexation. In the mixed aqueous solution of γ-CD and α-CD, the polymer guest is threaded with not only γ-CD but also α-CD (Scheme 7.5).[26] The end-capping reaction is performed with sufficiently bulky stoppers for α-CD, such as a trinitrophenyl group. When the threaded γ-CDs come to the ends of backbone polymers, the γ-CDs unthread over the stoppers in order. Once the first α-CD comes to the end, the unthreading of the CDs is stemmed. The end groups of the backbone prevent the α-CD from unthreading, and the stemmed α-CD blocks the subsequent γ-CDs. In this way, smaller CDs and their stoppers can stem larger CDs. However, this strategy is valid only for relatively less bulky guest polymers, and thus it can diversify the size of threaded CDs, but not the bulkiness of polymers.

To obtain polyrotaxanes that consist of diverse bulky polymers and γ-CD, the most promising capping groups are probably CDs themselves. The bulkiness of the smallest CD is even much larger than the cavity size of γ-CD, and the solubility in polar solvents is relatively good without a strong tendency for aggregation. Functionalized CDs, such as amino-modified ones, are available as capping regents.[27] Because such CD derivatives are soluble in water, the end-capping reaction can be conducted subsequent to the complexation process, carried out also in water (Scheme 7.6). As a result of the one-pot synthesis, γ-CD polyrotaxanes are obtained, though the efficiency of the synthesis is quite low. The reason for this is probably due to the low efficiency of the condensation reaction. This method has also the advantage of minimizing the dissociation of complexes during the capping reaction, because the reaction solvent is the same as that for the complexation. It is noteworthy that such CD derivatives are commercially available but very expensive.

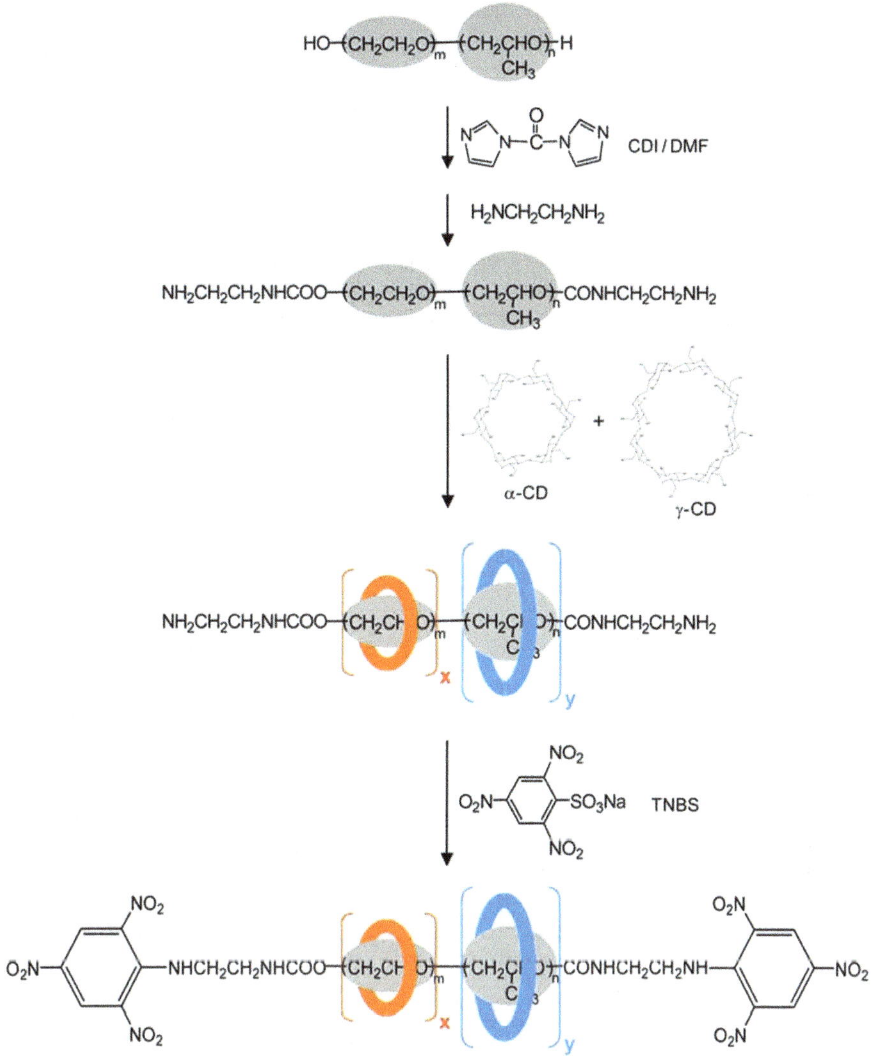

Scheme 7.5 Synthesis of a polyrotaxane composed of PEG–*ran*–PPG and an α-CD/ γ-CD mixture. The end groups are large enough to stem α-CDs, and the α-CDs hold back the γ-CDs (CDI, carbonyldiimidazole; TNBS trinitrobenzene sulfonate).[26]
Reprinted from *Polymer*, **49**, 4489–4493, Copyright (2008), with permission from Elsevier.

Unmodified CDs can be utilized when we carefully design the synthetic scheme to react the hydroxyl groups with the ends of backbone polymers. This means that the CDs that work as the cyclic component of pseudopolyrotaxane can be utilized also as the end-capping reagents. Both ends of guest polymers should be activated in advance to react with CDs. Active esters, such as *p*-nitrophenyl ester, can be employed. An example of

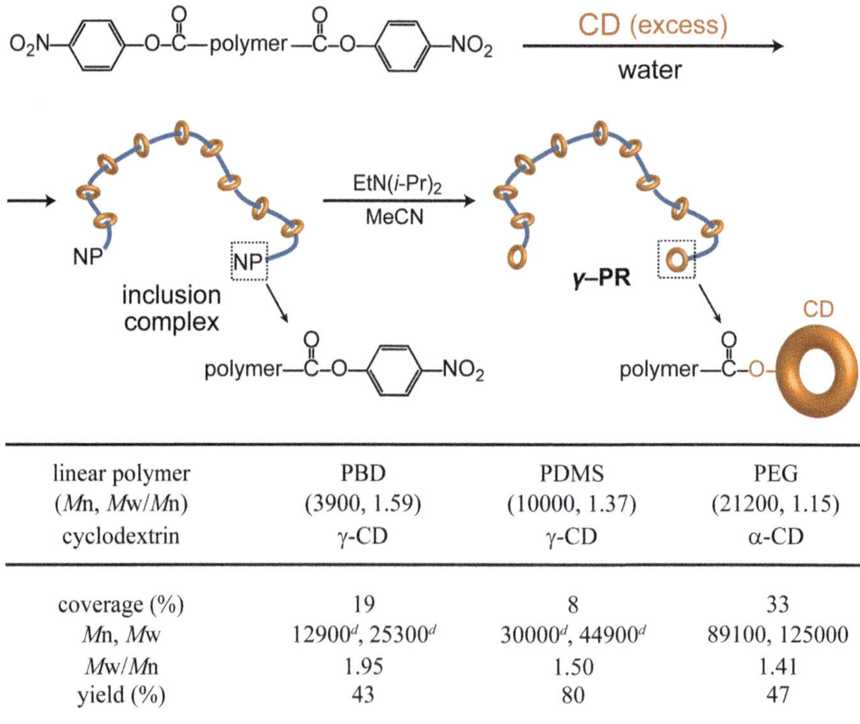

Scheme 7.6   Synthesis scheme of a PEG/γ-CD polyrotaxane by end-capping with an amino-modified CD as a stopper (DMT-MM, 4-(4,6-dimethoxy[1,3,5]-triazin-2-yl)-4-methylmorpholinium chloride).

linear polymer	PBD	PDMS	PEG
(Mn, Mw/Mn)	(3900, 1.59)	(10000, 1.37)	(21200, 1.15)
cyclodextrin	γ-CD	γ-CD	α-CD
coverage (%)	19	8	33
Mn, Mw	12900^d, 25300^d	30000^d, 44900^d	89100, 125000
Mw/Mn	1.95	1.50	1.41
yield (%)	43	80	47

Figure 7.18   Versatile polyrotaxane synthesis from different polymers and CDs by transesterification between the active ester at the polymer ends and hydroxyl groups of CDs.[28]
Reprinted with permission from *Macromolecules*. Copyright (2010) American Chemical Society.

this method to obtain a polyrotaxane that consists of polybutadiene (PBD) and γ-CD is introduced here (Figure 7.18).[28] Carboxyl-terminated PBD can be reacted with *p*-nitrophenol in the presence of carbodiimide to convert the end groups into a *p*-nitrophenyl ester. The activated polymer forms an inclusion complex with γ-CD in water. Only adding a tertiary amine, such as trimethylamine, to the complex can generate polyrotaxanes. It is notable that this method requires no additional capping reagents, because the cyclic component plays a double role as the cyclic host molecule and the capping reagent. The end-capping is achieved by the transesterification between the active ester and the ester formed with the hydroxyl group of the CD. Although the same reaction can occur between the active ester and the CDs threaded to dimerize the pseudopolyrotaxane, there are generally many free (non-threaded) CDs, because excess amount of CDs are used to accelerate the inclusion complexation. Therefore, such dimerization is generally negligible. Because additional capping regents that should be prepared for each inclusion complex are not necessary, this method is applicable for the synthesis of diverse polyrotaxanes. The same reaction scheme was applied for three different inclusion complexes, and then each polyrotaxane was obtained with relatively high yield. Polyrotaxane with polybutadiene as the backbone was firstly synthesized by this method. In addition, this method is applicable for a size-mismatched polyrotaxane that consists of PEG and γ-CD. These results prove that the capping method based on transesterification between the pre-modified end groups of the backbone and the excess cyclic components is very versatile.

7.4.2.2 Polymers Unrecognized by CDs

On the other hand, there are many polymers that form no inclusion complex with CDs. Poly(vinyl alcohol) is known to form no inclusion complex because of the high hydrophilicity. Polyolefins, such as polyethylene, are hardly captured by CDs. Even if such polymers could be forced to form the complexes by a change of process from the spontaneous complexation in water driven by the hydrophobic interactions, the subsequent end-capping requires functionalization of the end groups. It is not very easy to functionalize both ends of polyolefins. Moreover, it seems that there is no chance to synthesize polyrotaxanes with polymers that are thicker than the cavity of CDs [imagine, for example, a polyrotaxane that consists of poly(butyl methacrylate) and α-CD]. Is it really impossible to synthesize polyrotaxanes, if the backbone is not recognized by CDs? It is possible to synthesize not only such polyrotaxanes but also diversify the backbones of polyrotaxanes by using the following method.

Figure 7.19 shows the synthetic scheme of polyolefin-based polyrotaxanes.[29] Such polyrotaxanes cannot be synthesized directly from polyolefins because of the difficulties in the complexation with CDs and/or in the end-capping. However, we know that polyolefins can be obtained by addition reactions to polydienes; for example, the hydrogenation of 1,4-polybutadiene yields polyethylene. Such addition reactions can be applied for polyrotaxanes composed of polydienes. As mentioned previously, the polybutadiene-based polyrotaxane is now available because of the recent diversification of the backbones. A polyrotaxane that consists of polybutadiene and γ-CD

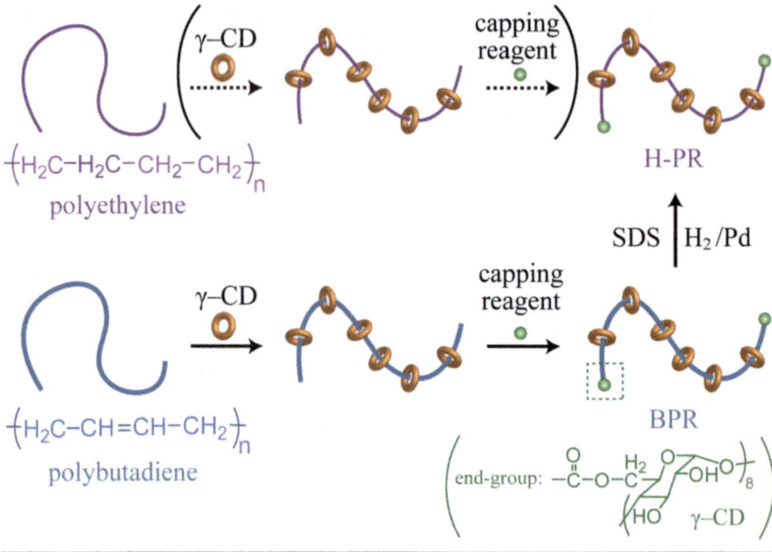

temp.	time (h)	degree of hydrogenation		yield (%)
		vinylene (%)	vinyl (%)	
RT	9	30	34	84
343 K	3	51	57	50
343 K	5	59	64	43
343 K	9	71	80	33
343 K	18	85	100	17
343 K	9	54	55	67

Figure 7.19   Synthesis scheme of polyolefin-based polyrotaxanes by *in-situ* modification of the backbone of a polybutadiene-based one, and the reaction conditions for different modification degrees of the unsaturated hydrocarbon.[29]
Reprinted from *Polymer*, **55**, 1514–1519, Copyright (2014), with permission from Elsevier.

synthesized by the above-mentioned method was hydrogenated with palladium black by the conditions shown in the table in Figure 7.19. The hydrogenation was progressed in water in the presence of a surfactant (sodium dodecyl sulfate, SDS). Although the polyrotaxane is not soluble in water, the surfactant forms a micelle with the polyrotaxane because of the amphiphilic nature of the polyrotaxane, which consists of a hydrophobic backbone polymer and hydrophilic rings. The micelle formation involving the polyrotaxane was also suggested by the fact that the hydrogenation degree is higher in the polyrotaxane compared to that in polybutadiene itself. Although the polybutadiene is located inside of the micelle, the polyrotaxane can be near the surface of the micelle because of the hydrophilic rings. Thus, the micelle formation involving the polyrotaxane has the advantage of both the increased possibility of encountering the reaction site, which is the surface of the catalyst, and the improved dispersity. As a result, the backbone polymer is converted into another one, while maintaining the interlocked structure of polyrotaxane.

This *in-situ* modification of the polyrotaxane backbone has an impact on not only the synthesis of polyrotaxanes that consist of unrecognized backbones by CDs, but also the diversification of the backbones of polyrotaxanes. Note that the modification degree is tunable by the reaction conditions, similar to other chemical modifications to conventional polymers. Various polyrotaxanes that have different modification degrees can be obtained from a single polyrotaxane. Theoretically, all chemical reactions that are available for polymers are applicable to polyrotaxanes that have those polymers as the backbones, though the reactivity for the polyrotaxanes are suppressed by the steric hindrance of CDs. Required performances for the polyrotaxanes are the reactivity of the backbones and the chemical stability of the rotaxane structure. From this aspect, the unsaturated carbon double bonds are ideal, because of the variety of addition reactions to the double bonds without cleavage of the main chain. In addition, the design of end-capping reactions should also be considered carefully. For example, when the capping groups are bound to the backbone *via* an ester bond, which is subject to hydrolysis, the subsequent modifications to the backbones are limited under mild conditions without strong acids, bases, oxidation, and reducing agents.

Converting the backbone of polyrotaxane has a great impact on the morphology. As an example, variation of the crystal structures in a polyrotaxane induced by the hydrogenation of the polybutadiene-based polyrotaxane is introduced here. Polyrotaxanes that have different hydrogenation degrees were synthesized by the above-mentioned method. The molecular weight of the backbone is $M_\mathrm{n} = 3900$ and $M_\mathrm{w} = 6200$, and the vinylene/vinyl unsaturation distribution is 79/21. Profiles of power X-ray diffraction of these polyrotaxanes are shown in Figure 7.20. As described in Section 7.3, CD-based polyrotaxanes generally form hexagonal columnar crystals due to the efficient hydrogen bonds between threaded CDs. The reactant polybutadiene-based polyrotaxane also exhibits a typical columnar crystal structure of the threaded γ-CDs with the characteristic peaks of (220) and (420) crystal faces. Incidentally, the strong peak of the (420) face is usually regarded as an indicator for the inclusion complexation with γ-CD. The peaks of the CD's crystal are gradually weakened with the increase of the hydrogenation degree. At the same time, new peaks are generated that correspond to the (110) and (200) faces of the orthorhombic crystal structure of polyethylene. The increase of these new peaks with increasing hydrogenation degree indicates the crystal formation of the backbone polymer of the polyrotaxane induced by the reaction. These results suggest the morphological change of the polyrotaxane in the solid state by the conversion of the backbone from the amorphous polydiene into the crystalline polyolefin, as illustrated in Figure 7.21. The tendency of the crystal formation of

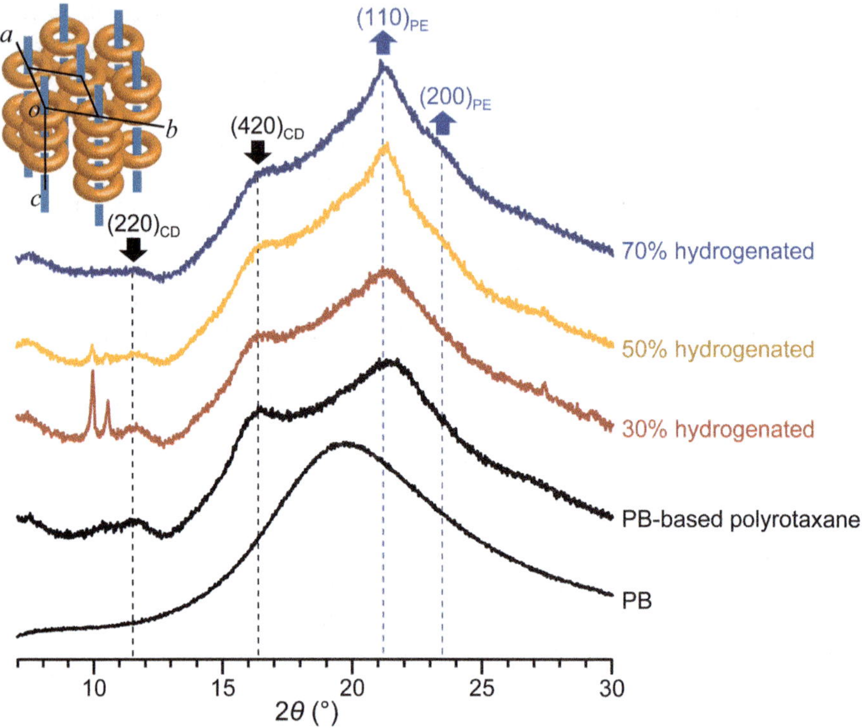

Figure 7.20 X-Ray diffraction profiles of a polybutadiene-based polyrotaxane and the derivatives with different degrees of hydrogenation of the backbone. Peaks that correspond to (110) and (200) planes of polyethylene generated with hydrogenation are marked.[29]
Reprinted from *Polymer*, **55**, 1514–1519, Copyright (2014), with permission from Elsevier.

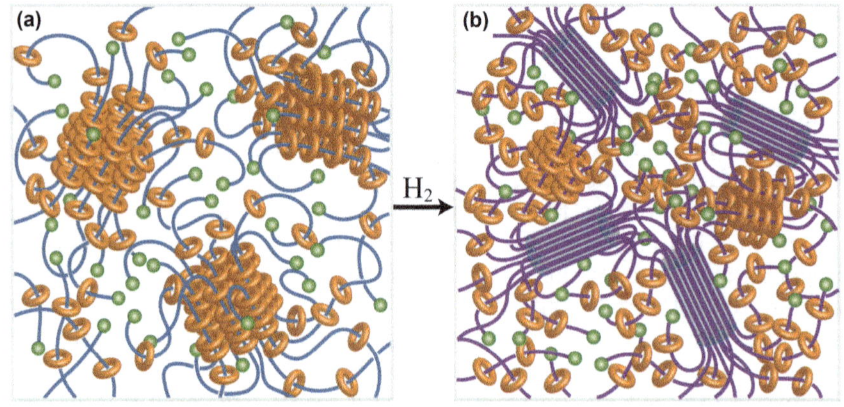

Figure 7.21 Morphological transition of polyrotaxane. Columnar crystal formation by threaded CDs in polybutadiene-based polyrotaxane (left) is suppressed by hydrogenation of the backbone polymers to form alternatively crystals of the backbone (right).[29]
Reprinted from *Polymer*, **55**, 1514–1519, Copyright (2014), with permission from Elsevier.

polyethylene [poly(ethylene-butylene) accurately] seems to be strong enough to push aside the ring components. Once the backbone forms the crystal, CDs cannot form their crystals at the same place, meaning that the hydrogenated backbone and threaded CDs are competing to form their crystals. Therefore, the morphology varies with the modification degree, and the degrees are easily tunable by control of the reaction conditions for the *in-situ* modification.

One may think that it is also possible to synthesize such a variety of polyrotaxanes with different modification degrees from the corresponding pre-modified polymers by using the conventional threading methods, as long as the polymers have reactive end groups for capping. However, the optimal conditions for the inclusion complexation and the end-capping will be changed considerably according to the modification degree. The most serious problem in the one-by-one synthesis is the variable structures of the inclusion complexes. Even if we can manage to find out the optimal conditions for the complexation and capping, it is probably impossible to control the stoichiometry of the guest polymers and CDs. Thus, the *in-situ* modification process has a great advantage for the synthesis of a series of polyrotaxanes with different backbones and the same coverage.

7.4.3 Diversity of the Host–Guest Ratio (Including Double-threaded PR)

The stoichiometry of the guest polymer and the host CDs is crucial for various properties of polyrotaxanes and slide-ring materials. Nevertheless, the control of the coverage of the polyrotaxane has not been developed very much. The main reason of the slow development is that the stoichiometry has not been considered as an important factor to determine the properties of polyrotaxanes and slide-ring gels. Of course, whether the backbone polymer is covered with CDs sparsely or densely has been an important issue. Densely covered polyrotaxanes have been utilized for molecular insulated wires; conductive polymers used as the backbone polymers can be insulated by the inclusion with CDs.[30] Thus, many efforts have been devoted to increase the density of the CDs on the backbone polymers. Such dense polyrotaxanes are also required to synthesize a molecular tube composed of CDs. The molecular tube is obtained by binding neighboring CDs in a single polyrotaxane, as shown in Scheme 7.7.[31] Thus, it is regarded that the coverage of the polyrotaxane should be higher to realize the ideal reaction to obtain the nanotube. On the other hand, sparse polyrotaxanes are also required to utilize the intramolecular motions such as sliding and rotation. As mentioned in Section 7.4.1, many derivatives of polyrotaxanes for their applications have been synthesized from sparse polyrotaxanes. For slide-ring materials, such sparse polyrotaxanes are essential as well to enable the characteristic chain sliding through the cross-links.

In this way, the importance of the stoichiometry between CDs and the polymer guest, which can be called the density of the CDs in polyrotaxanes,

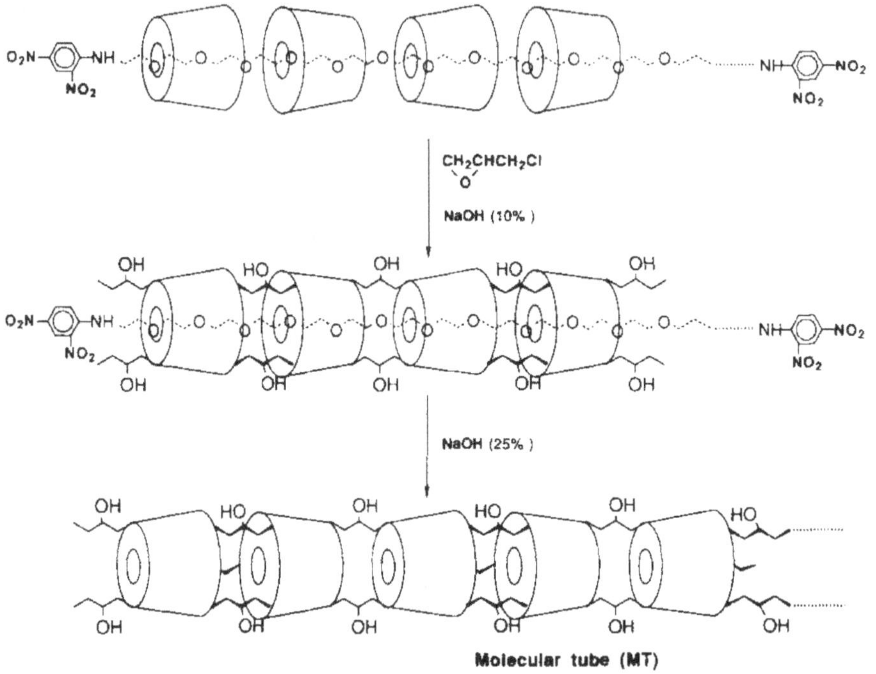

$$CH_2CHCH_2Cl$$

NaOH (10%)

NaOH (25%)

Molecular tube (MT)

Scheme 7.7   Synthesis scheme of a nanotube from a densely covered polyrotaxane.[31] Reprinted by permission from Macmillan Publishers Ltd: *Nature*, **367**, 516–518 copyright (1993).

has been recognized very roughly; the single most important issue was whether it was dense or sparse. One may be satisfied with a sparse polyrotaxane, as long as the ability of the sliding appears to be guaranteed. When the coverage is 25%, which means that three quarters of the main chain is naked, one may think that the sparse polyrotaxane has enough room for the sliding of CDs. However, as mentioned in Section 4.3, the ability of sliding can be limited by the entropic contribution of the ring components. Recall that the threaded CDs can behave like gas molecules confined in a cylinder, and the cross-linking of polyrotaxane can be a way to extract the entropy of the CDs on a molecular level into the macroscopic properties of slide-ring materials. Thus, even though most of the surface of a guest polymer is naked, sliding is not guaranteed and is suppressed by the "pressure" of threaded CDs. In addition, it is expected that the mechanical properties of slide-ring materials can be controlled by the density of CDs in polyrotaxanes.

Before the entropic contribution was proposed, reducing the density of CDs in polyrotaxane had been tried with the expectation of more efficient sliding due to the provision of larger space. Sparse polyrotaxanes are obtained from relatively long guest polymers. Short PEGs with $M_W < 3000$ form dense polyrotaxanes with α-CDs, whereas longer PEGs tend to form

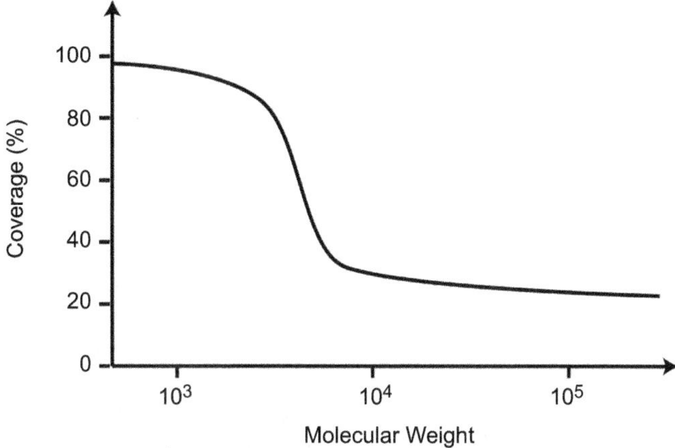

Figure 7.22   Outline of the coverage of a PEG/α-CD polyrotaxane depending on the molecular weight of PEG.

more sparse polyrotaxanes. However, the coverage becomes almost constant at 20–30% with $M_W > 10\,000$. The rough dependence of the coverage of PEG/α-CD polyrotaxanes on the molecular weight of PEG is roughly drawn in Figure 7.22. This tendency clearly shows that it is only possible to select sparse or dense by the molecular weights of the guest polymers, but it is impossible to obtain much sparser ones.

To reduce the coverage of polyrotaxanes that are spontaneously determined by molecule recognition, several methods are available. Promotion of dissociation of the inclusion complex can reduce the coverage.[32] Selection of the reaction solvent for end-capping that promotes the dissociation can reduce the coverage of the polyrotaxane. For example, mixed solvents between DMSO and DMF can be used for the pseudopolyrotaxanes of PEG/α-CD. DMF is a frequently used solvent for the end-capping, because it is a good solvent for various capping reagents but not for the pseudopolyrotaxanes (recall the selection of solvents described in Section 7.2.3). On the other hand, DMSO is the most commonly used solvent for CD-based polyrotaxanes, because the hydrogen bonds between threaded CDs can be intercepted by the solvent (see Section 7.2.4). Since good solvents for polyrotaxanes have the ability to dissociate threaded CDs, these solvents can induce the dethreading of CDs from the inclusion complexes. The end-capping reaction is carried out with the promotion of dethreading. As a result, the number of CDs entrapped is decreased. The coverage of polyrotaxane synthesized by the end-capping in mixed solvents decreases with an increase of the fraction of DMSO (Table 7.3). However, the yield of polyrotaxane is also decreased, because this method decreases the coverage by promoting the dissociation of the inclusion complex.

Another strategy is in the inhibition of the inclusion complexation. The complexation reaction at higher temperature can reduce the number of threaded CDs,[33] because the inclusion process is an entropically unfavorable

Table 7.3 Coverages and yields of PEG/α-CD polyrotaxanes synthesized by end-capping reactions in mixed solvents.[32] Reproduced by permission of John Wiley & Sons, Inc.

Reaction solvent (DMF/DMSO)	Coverage (%)	Yield (%)
100/0	53	28
90/10	37	26
85/15	22	6

process. In the case of the PEG/α-CD system, precipitates of the inclusion complexes are generated from their solutions in water at room temperature (see Section 7.2.2). No precipitation occurs at 70 °C, and slightly turbid solution is obtained at 35 °C. These appearances indicate no or slight complexation. Incidentally, precipitates are generated after cooling these solutions below room temperature. By the subsequent end-capping in water to the halfway complex at 35 °C, polyrotaxanes with very low coverage are obtained. However, the significant reduction of the yield is unavoidable because of the incomplete complexation.

These methods to control the coverage are based on modifications of the reaction process. Originally, the control of the coverage is the issue of host–guest chemistry. As known in traditional host–guest chemistry with CDs, various molecular designs have been investigated for controlling the host–guest stoichiometry. Thus, another approach based on the molecular design of the host and guest molecules should be effective also for controlling the density of CDs in polyrotaxanes. Modifications of CDs usually result in no complexation with guest polymers, because the hydroxyl groups of CDs are necessary to form the inclusion complexes efficiently. On the other hand, the end groups of the guest polymers have a great impact on the stoichiometry of the complexes. It has been known that the bulkiness of the end groups influence the kinetics of the complexation with CDs.[34] For example, the kinetics of the complexation between α-CD and linear hydrocarbons is significantly affected by the number and position of methyl substituents on the phenyl end groups (Figure 7.23). In these cases, the possible stoichiometry of the host and guest is only 1 : 1 at the equilibrium state, though the kinetics are different. Although the host–guest ratio is diverse in the case of polymer guests (1 : 1, 2 : 1, 3 : 1, . . .), the same effect of the end groups on the complexation kinetics is expected as well.

However, not only the complexation kinetics but also the stoichiometry is drastically affected by the introduction of the bulky end groups to the guest polymers.[35] The number of α-CD threaded with PEG is drastically reduced by modification of the hydroxyl end groups of PEG with 2,2-dimethylsuccinic anhydride (Table 7.4). Since the synthetic procedures are essentially the same as the conventional method starting from carboxyl-terminated PEG, the difference in the coverage arises from the end groups. Notably, the yield is considerably high compared to the above-mentioned syntheses based on

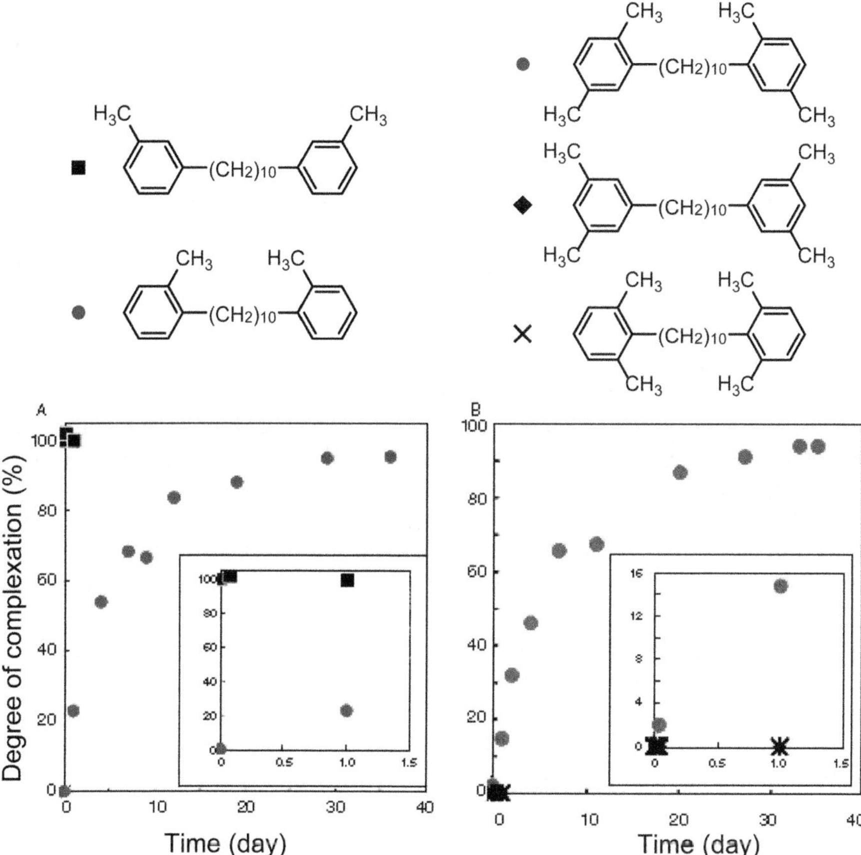

Figure 7.23 Time course of the complexation between α-CD and the hydrocarbon guest molecules with (A) mono-substituted and (B) di-substituted phenyl end groups.[34]
Reproduced by permission of IOP Publishing.

Table 7.4 Coverages and yields of PEG/α-CD polyrotaxanes synthesized by end-capping reactions in mixed solvents.[35]

End group of PEG	Coverage (%)	Yield (%)
$-CO_2H$	25	$>90^a$
$-CH_2-C(CH_3)_2-CO_2H$	5	69

[a]From J. Araki *et al.*, *Macromolecules*, 2005, **38**, 7524–7527 and repeated additional tests.

the promotion of dethreading or the incomplete complexation. The significantly lowered coverage is reproducible within the margin of error: $5.1 \pm 0.5\%$ $(n = 7)$. The high yield and reproducibility clearly indicates that the introduced end groups drastically change the equilibrium of the complexation, not only the kinetics. Although the mechanism of the inclusion complexation itself is still not clear, the introduction of such bulky end groups to guest polymers seems to be a promising way to control the coverage of polyrotaxanes.

The reduction of coverage significantly contributes to the drastic improvement of the mechanical properties of the slide-ring materials. The obtained polyrotaxane with 5% coverage is cross-linked in DMSO to form slide-ring gels. A conventional polyrotaxane that has the same PEG with the same molecular weight and different coverage (25%) is also cross-linked. The only difference is the density of the ring components of the slide-ring gels. Surprisingly, the extensibility of the slide-ring gels with lower coverage is much higher than that of gels with higher coverage. Such drastic improvement of the extensibility cannot be explained only by the increased room for sliding. As mentioned in Section 4.3, the sliding ability can be predicted to compete the entropic force of the ring components. Thus, the obvious difference in extensibility is attributable to the considerable difference in the entropy of the rings. The rings confined between the cross-links can behave as gas molecules, and thus the "pressure" can increase with the density of the rings. Because the ring density is considerably decreased, the chain sliding occurs efficiently without inhibition by the ring components to exhibit high extensibility.

The stoichiometry of the guest polymer and CDs becomes more diverse when double-stranded polyrotaxanes are possible. Such polyrotaxanes are only possible with γ-CD or larger ones, and the characterization is complicated, as mentioned in Section 7.3. Although there are many reports that various thin polymers can easily form such double-stranded complexes with γ-CD, there are at least two requisites for the guest polymers. One is that the molecular weight should be relatively short. The first report was a double-stranded complex between PEG and γ-CD, and the molecular weight of PEG was 1500. As mentioned previously, such short PEG chains tend to form densely packed polyrotaxanes. Indeed, the coverage of the inclusion complex was almost 100%. Because the entropic loss of the guest polymer increases with the molecular weight, such a double-stranded complex is hardly formed with long PEGs. Another is suitable functional groups at the ends of polymers for the complexation. As shown in Figure 7.14, the first example of the double-stranded one was obtained from PEG with aromatic end groups, and the complex was also produced from PEG with various aromatic groups. These results suggest that the interactions between the end groups of guest polymers are essential for double-stranded complexes, as mentioned previously. Indeed, a single-stranded polyrotaxane (not pseudopolyrotaxane) was obtained from γ-CD and PEG ($M_n = 2000$) without such aromatic end groups.[36] Even when PEG has aromatic end groups, long PEGs with $M_n = 20\,000$ yield single-stranded ones.[37] Because these polyrotaxanes consist of a thin polymer and large CDs, the cavity of threaded γ-CD has extra rooms for other guests (Figure 7.24). Inclusion of solvent molecules into the extra cavity has a considerable impact on the macroscopic dynamics of slide-ring gels as described in Section 4.5.[38]

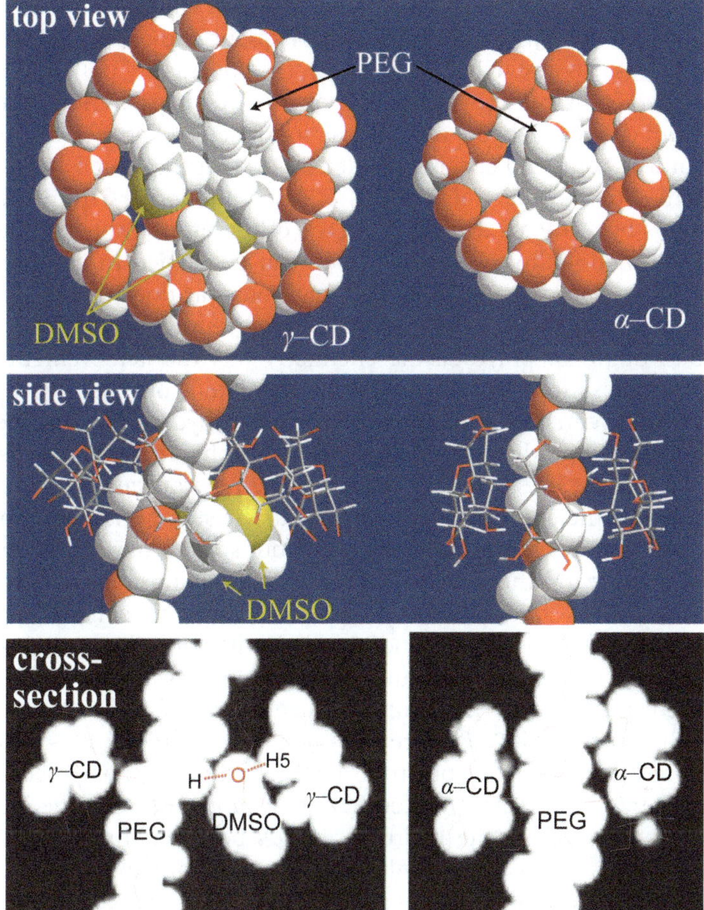

Figure 7.24   Molecular models of the PEG/α-CD and PEG/γ-CD polyrotaxane. The cavity of γ-CD has an extra cavity for inclusion of solvent molecules (DMSO). Reproduced from Ref. 37 by permission of The Royal Society of Chemistry.

7.5   Simplification of the Threading Method

As described so far, each elemental step of the threading method has been progressed, and the chemical structures of CD-based polyrotaxanes has been diversified. Although the individual steps are currently getting more sophisticated and diversified, the synthesis of polyrotaxane is normally still step-by-step, following the four steps. Here, we introduce recent examples that can simplify the threading method.

The main strategy for the simplification is reducing the number of steps. Recall that the inclusion complexation process should be processed in water because of the hydrogen bonds that progress the complexation. Activation of the end groups of guest polymers and the end-capping reaction is usually performed in organic solvents. Thus, after activation, the organic solvent should be exchanged into water for the subsequent complexation, and then the water should be eliminated for the subsequent end-capping. If the end-capping process can be carried out in water, the "dormant" polyrotaxane can be synthesized from the activated guest polymer in one pot. Imine formation from an amine and aldehyde, a condensation reaction between a carboxylic acid and amine-mediated specific condensation reagents, such as DMT-MM (see Scheme 7.6), and polymerizations of bulky monomers from the guest polymer ends by ATRP are available. On the other hand, if the complexation can be performed without water, all steps can be carried in one pot, theoretically. This means a completely new method that does not rely on self-assembly between CDs and the guest polymers. Remember that the historical breakthrough for polyrotaxane chemistry was the achievement of the efficient complexation by self-assembly mediated by the hydrophobic interactions between CDs and guest polymers and hydrogen bonds between threaded CDs. Similarly, a breakthrough of such efficient complexation without water must bring a great impact to the polyrotaxane synthesis.

Although the efficiency is not comparable to the step-by-step method thus far, some examples of the complexation without solvent have been reported. A scheme of one of the finest syntheses of polyrotaxanes is shown in Scheme 7.8.[38] The solid of methylated α-CD and the solid of polyTHF ($M_n = 1000 \sim 2900$) were mixed and grinded without solvent. Subsequent end-capping by just adding the liquid of a bulky isocyanate with a catalyst yields the polyrotaxane. Polyrotaxanes with coverages of $28 \sim 81\%$, which are higher with shorter polyTHF, were obtained in good yield: $26 \sim 44\%$. Notably, the product is an active polyrotaxane, not a dormant one, because the hydroxyl groups of the CD were modified in advance. Because the inclusion process does not rely on the hydrogen bonds between threaded CDs, the activation process is not needed. Although the mechanism of complexation is still unclear, the improvement of such a solvent-free inclusion process will bring innovation in the application of polyrotaxanes, as this example indicated.

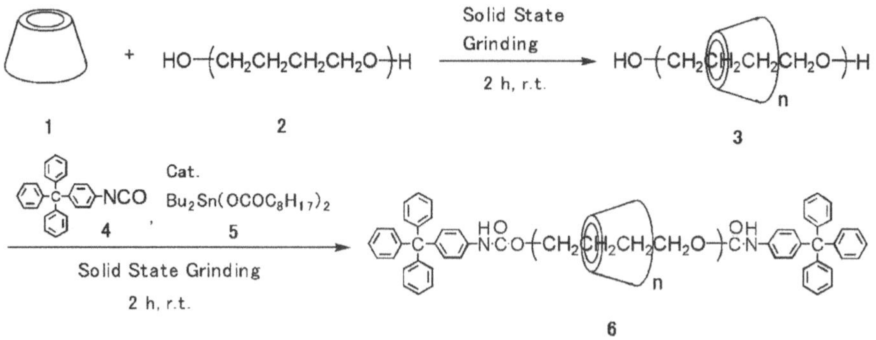

Scheme 7.8   Solid-state synthesis of a polyrotaxane that consists of O-trimethyl-α-CD (1) and poly(tetrahydrofuran) ($M_n = 1000 \sim 2900$) (2).[38] Reproduced by permission of John Wiley & Sons, Inc.

We introduce here an example of the simplification of the synthesis of graft-polyrotaxane, which is industrially very important currently (see Section 7.4.1 and Section 8.4). Recall that graft-polyrotaxanes can be synthesized from activated polyrotaxanes, not dormant ones, because the grafting polymerization from unmodified threaded CDs usually results in the production of insoluble cross-linked solids. Thus, another step of the grafting should be added to the four steps of the polyrotaxane synthesis. The conventional synthesis that consists of five steps can be simplified by performing the end-capping, the activation, and the grafting at once.[39] The scheme of the three-step synthesis for a graft polyrotaxane is shown in Scheme 7.9. Activation of the end groups of PEG was achieved by the addition of gluconic acid. Then, the polymer forms an inclusion complex with α-CD. Finally, the obtained pseudopolyrotaxane was mixed with ε-caprolactone and a catalyst, and then the mixture was heated to initiate the polymerization of the monomer. The point is that the ring-opening polymerization of ε-caprolactone can start not only from the hydroxyl groups of threaded CDs but also from those at the ends of backbone polymers. At the same time, the base catalyst, 1,8-diazabicycloundec-7-ene (DBU), plays a double role as a polymerization catalyst and an activation reagent that dissociates the hydrogen bonds between threaded CDs. The dethreading of the CDs from the backbone can be tuned by the temperature; too high temperature for the reaction induces rapid dissociation of the complex, and too low temperature results in too slow end-capping to trap CDs. A kinetically controlled reaction can achieve a relatively high yield of the graft-polyrotaxane: 33～69%.

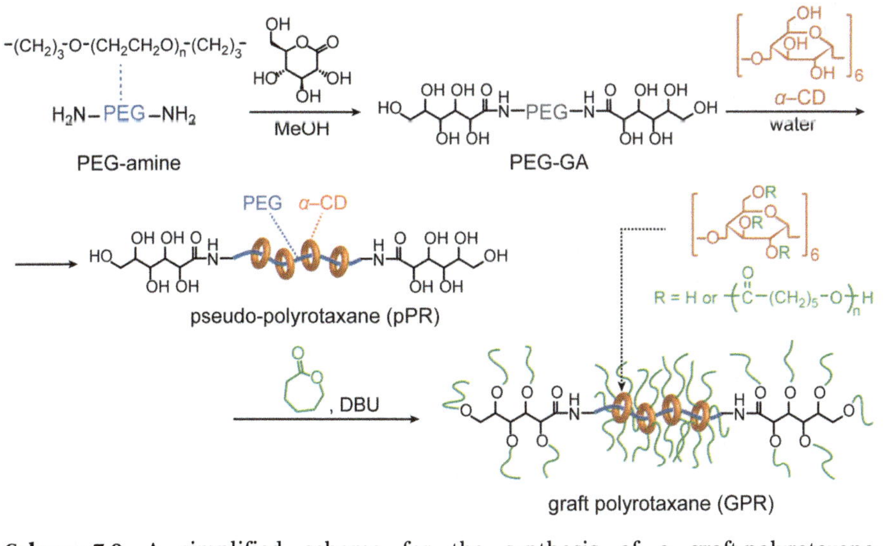

Scheme 7.9 A simplified scheme for the synthesis of a graft-polyrotaxane that consists of a PEG backbone, α-CD, and graft chains of poly(ε-caprolactone).[39]

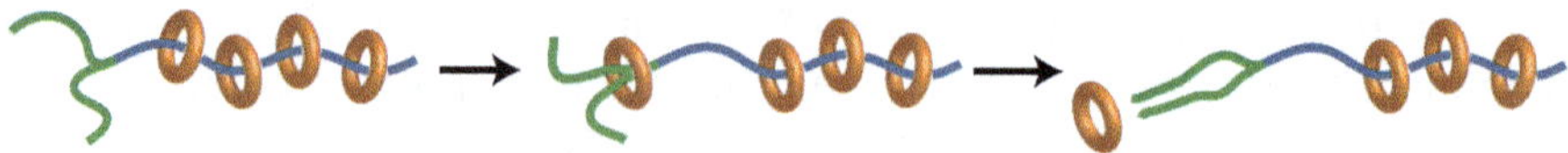

Scheme 7.10 Scheme of the dethreading process in the presence of a branch structure on the backbone polymer. Threaded CDs can be dissociated through the double-stranded state (middle).

One may notice that poly(ε-caprolactone) is a thick polymer that can form an inclusion complex with α-CD (see Table 7.1). How can such thin polymer prevent the dethreading of CDs? The point is the branch of the backbone chain at the ends. When a single poly(ε-caprolactone) chain is attached at each end of the polymer, the threaded α-CDs are completely dethreaded experimentally, indicating that the chain is certainly too thin to trap α-CD. On the other hand, when the multiple chains are grafted at each end of the backbone polymer, α-CD can be trapped. To escape from the backbone, the threaded CDs must overcome the double-threaded state, as schematically shown in Scheme 7.10. The cavity of α-CD is too small to allow any polymer chains to penetrate with another chain simultaneously. This kind of end-capping by introducing branch structures at the backbone polymer is fundamentally different from the concept of conventional end-capping, as described in Section 7.2.3, which is based on the introduction of a bulky group at each end. It is notable that this new concept for end-capping is theoretically valid for any inclusion complex to form polyrotaxanes, particularly for the troublesome synthesis of γ-CD-containing polyrotaxanes.

References

1. L. Fang, M. A. Olson, D. Benítez, E. Tkatchouk, W. A. Goddard III and J. F. Stoddart, *Chem. Soc. Rev.*, 2010, **39**, 17–29.
2. K. S. Chichak, S. J. Cantrill, A. R. Pease, S. H. Chiu, G. W. V. Cave, J. L. Atwood and J. F. Stoddart, *Science*, 2004, **304**, 1308–1312.
3. F. M. Raymo and J. F. Stoddart, *Pure Appl. Chem.*, 1997, **69**, 1987–1997.
4. F. Biedermann, W. M. Nau and H.-J. Schneider, *Angew. Chem., Int. Ed.*, 2014, **53**, 11158–11171.
5. Y. Okumura, K. Ito and R. Hayakawa, *Polym. Adv. Technol.*, 2000, **11**, 815–819.
6. A. Harada, A. Hashidzume, H. Yamaguchi and Y. Takashima, *Chem. Rev.*, 2009, **109**, 5974–6023.
7. A. Ueno, F. Moriwaki, T. Osa, F. Hamada and K. Murai, *J. Am. Chem. Soc.*, 1988, **110**, 4323–4328.
8. M. Arunachalam and H. W. Gibson, *Prog. Polym. Sci.*, 2014, **39**, 1043–1073.
9. T. Oshikiri, Y. Takashima, H. Yamaguchi and A. Harada, *J. Am. Chem. Soc.*, 2005, **127**, 12186–12187.
10. A. Harada, J. Li and M. Kamachi, *Nature*, 1992, **356**, 325–327.

11. K. Kato, K. Inoue, M. Kidowaki and K. Ito, *Macromolecules*, 2009, **42**, 7129–7136.
12. Y. Sakai, K. Ueda, N. Katsuyama, K. Shimizu, S. Sato, J. Kuroiwa, J. Araki, A. Teramoto, K. Abe, H. Yokoyama and K. Ito, *J. Phys.: Condens. Matter*, 2011, **23**, 284108.
13. A. Takahashi, R. Katoono and N. Yui, *Macromolecules*, 2009, **42**, 8587–8589.
14. A. Harada, J. Li, T. Nakamitsu and M. Kamachi, *J. Org. Chem.*, 1993, **58**, 7524–7528.
15. A. Harada, J. Li and M. Kamachi, *Nature*, 1994, **370**, 126–128.
16. J. Araki and K. Ito, *J. Polym. Sci. Part A: Polym. Chem.*, 2006, **44**, 6312–6323.
17. H. Hyun and N. Yui, *Macromol. Biosci.*, 2011, **11**, 765–771.
18. J. Araki, T. Kataoka and K. Ito, *Soft Matter*, 2008, **4**, 245–249.
19. C. Teuchert, C. Michel, F. Hausen, D. Park, H. W. Beckham and G. Wenz, *Macromolecules*, 2013, **46**, 2–7.
20. J. Seo and N. Yui, *Biomaterials*, 2013, **34**, 55–63.
21. K. Kato, D. Matsui, K. Mayumi and K. Ito. *submitted*.
22. K. Kato, K. Inoue, M. Kidowaki and K. Ito, *Macromolecules*, 2009, **42**, 7129–7136.
23. M. Okada, Y. Takashima and A. Harada, *Macromolecules*, 2004, **37**, 7075–7077.
24. J. W. Peng Gao, P. Wang, L. Ye, A. Zhang and Z. Feng, *Polymer*, 2011, **52**, 347–355.
25. J. Wang, P. Gao, L. Ye, A. Zhang and Z. Feng, *Polym. Chem.*, 2011, **2**, 931.
26. C. Yang, X. Ni and J. Li, *Polymer*, 2009, **50**, 4496–4504.
27. A. Takahashi, R. Katoono and N. Yui, *Macromolecules*, 2009, **42**, 8587–8589.
28. K. Kato, H. Komatsu and K. Ito, *Macromolecules*, 2010, **43**, 8799–8804.
29. K. Kato, T. Ise and K. Ito, *Polymer*, 2014, **55**, 1514–1519.
30. M. J. Frampton and H. L. Anderson, *Angew. Chem., Int. Ed.*, 2007, **46**, 1028–1064.
31. A. Harada, J. Li and M. Kamachi, *Nature*, 1994, **365**, 516–518.
32. T. Ooya, M. Eguchi and N. Yui, *J. Am. Chem. Soc.*, 2003, **125**, 13016–13017.
33. G. Fleury, C. Brochon, G. Schlatter, G. Bonnet, A. Lapp and G. Hadziioannou, *Soft Matter*, 2005, **1**, 378–385.
34. H. Yamaguchi, T. Oshikiri and A. Harada, *J. Phys.: Condens. Matter*, 2006, **18**, S1809–S1816.
35. K. Kato, Y. Okabe, Y. Okazumi, K. Ito, *submitted*.
36. A. Takahashi, R. Katoono and N. Yui, *Macromolecules*, 2009, **42**, 8587–8589.
37. K. Kato, K. Karube, N. Nakamura and K. Ito, *Polym. Chem.*, 2015, **6**, 2241–2248.
38. R. Liu, T. Maeda, N. Kihara, A. Harada and T. Takata, *J. Polym. Sci., Part A: Polym. Chem.*, 2007, **45**, 1571–1574.
39. K. Kato, K. Inoue, M. Kudo and K. Ito, *Beilstein J. Org. Chem.*, 2014, **10**, 2573–2579.

Applications of Polyrotaxane and Slide-ring Materials

8.1 Biological Applications of Polyrotaxane

Polyrotaxane has a topologically interlocked structure of ring and string components, which are not covalently bound to each other. The unique supramolecular structure is maintained by bulky stoppers at both ends of the backbone string. If the stopper is removed, rings are dissociated from the axis chain immediately. This drastic change of the unique structure can be applied for biological uses such as drug delivery, and so on. In practice, Yui and Ooya first developed α-cyclodextrin-based biodegradable polyrotaxanes where bulky stoppers were connected to the backbone poly(ethylene glycol) with an oligopeptide (Figure 8.1).[1] When the terminal residue can be cleaved by enzymatic hydrolysis *in vivo*, supramolecular dissociation is triggered by the terminal hydrolysis. If hydroxyl groups in α-cyclodextrins are chemically modified with drug molecules, dissociated α-cyclodextrins act effectively as drugs in the body. In addition, poly(ethylene glycol) and cyclodextrins are biocompatible, and cyclodextrins are approved by the US Food and Drug Administration (FDA). This means that biodegradable drug–polyrotaxane conjugates have biosafety and biocompatibility, and are useful as a drug delivery system.[2,3]

The enzymatic degradation of such biodegradable polyrotaxanes proceeds efficiently in lysosomes.[4] When papain was used as an enzyme for degradation, terminal hydrolysis in the polyrotaxane proceeded with a conversion rate of over 85%. In contrast, L-Phe-terminated poly(ethylene glycol) as a backbone string showed limited hydrolysis of about 50%. The difference was ascribed to the accessibility of papain to the terminal peptide linkage of the backbone poly(ethylene glycol), dependent on the polymer conformation. The specific association between papain and the rod-like drug–polyrotaxane

Monographs in Supramolecular Chemistry No. 15
Polyrotaxane and Slide-Ring Materials
By Kohzo Ito, Kazuaki Kato and Koichi Mayumi
© Kohzo Ito, Kazuaki Kato and Koichi Mayumi 2016
Published by the Royal Society of Chemistry, www.rsc.org

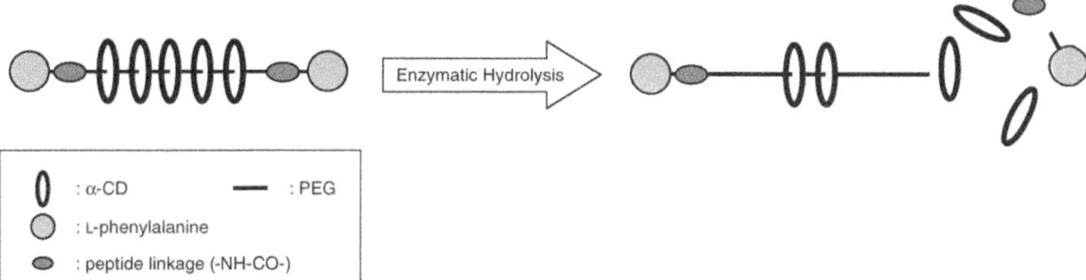

Figure 8.1   Schematic diagram of biodegradable polyrotaxanes of poly(ethylene glycol) (PEG) and α-cyclodextrins (α-CD).[2]
Reproduced by permission of The Japanese Society for Artificial Organs.

conjugate should expose the terminal linkages to the aqueous environment. The complete release of drug-functionalized α-cyclodextrins was achieved from the drug–polyrotaxane conjugate after about 200 to 320 h. Consequently, cyclodextrin dethreading from the backbone string by the enzymatic hydrolysis of the terminal linkage is quite promising for the possible applications to drug delivery systems such as cytoplasmic dissociation caused by lysosomal enzymes.

Another interesting feature of polyrotaxanes for biomedical use is to enhance the multivalent interaction between the ligand and receptor in ligand–conjugated polyrotaxanes.[2,3] When the ligand is covalently bound to an α-cyclodextrin, the ligand can move along the backbone poly(ethylene glycol) chain freely, together with α-cyclodextrin. This high mobility should enhance the multivalent binding of the ligand–conjugated polyrotaxane in cells and tissues. Yui *et al.* actually synthesized biotin–polyrotaxane conjugates, and investigated their interaction with streptavidin-immobilized surfaces by surface plasmon resonance (SPR) spectroscopy.[5] The equilibrium association constant indicated that the interaction between biotin and the streptavidin surface was enhanced with an increasing number of biotin molecules in the polyrotaxanes, suggesting a multivalent interaction.

When the ligand is an oligopeptide, the polyrotaxane conjugate shows an inhibitory effect on intestinal human peptide transporters (hPEPT1). It was reported that the oligopeptide–polyrotaxane conjugates had a much stronger inhibitory effect on the uptake of digested oligopeptides through hPEPT1 than that of oligopeptide–dextran conjugates, which was ascribed to the multivalent binding between the polyrotaxane conjugates and hPEPT1.[6] Since the oligopeptide–dextran conjugates have strong hydrophobic interactions among the oligopeptides, the steric hindrance may disturb the interaction between the oligopeptide and hPEPT1 on cell surfaces. On the other hand, polyrotaxanes have a rod-like structure to enhance the access of the oligopeptide to hPEPT1.

In addition, Yui and coworkers developed maltose–polyrotaxane conjugates for carbohydrate recognition.[7] Generally, the interaction between carbohydrates and their receptors is so weak that the specificity is not high. The maltose–polyrotaxane conjugates exhibited a >3000 times stronger

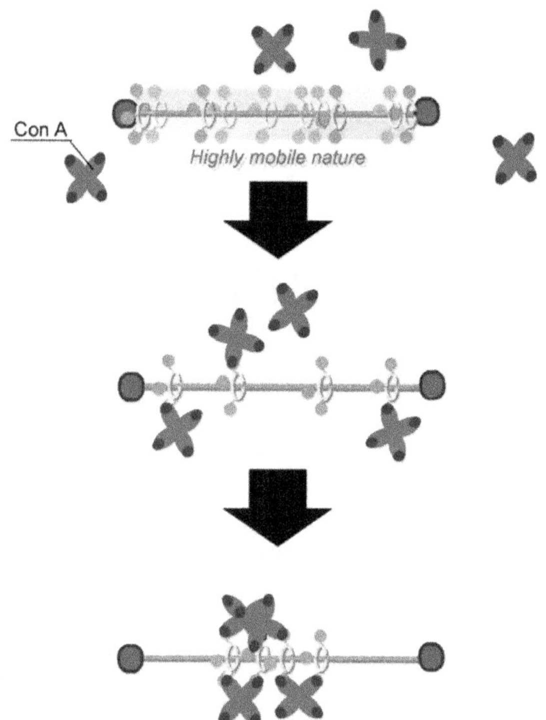

Figure 8.2 Illustration of ConA binding with maltose–polyrotaxane conjugates.[2] Reproduced by permission of The Japanese Society for Artificial Organs.

inhibitory effect on concanavalin A- (ConA)-induced hemagglutination as that of maltose, as shown in Figure 8.2. From the potency of the polyrotaxane molecular unit, the inhibitory effect was evaluated to be over 700 000 times that of a maltose molecule. The inhibitory effect is enhanced with an increasing number of maltose groups in the polyrotaxane, which suggests the multivalent interaction between maltose and ConA. Interestingly, an increasing number of α-cyclodextrin units diminished the inhibitory effect, while the inhibitory effect had no relation to the density of maltose groups per α-cyclodextrin. This was attributed to the rigidity of the polyrotaxanes. Nuclear magnetic resonance (NMR) analysis revealed that the high mobility of ligands with α-cyclodextrin along poly(ethylene glycol) enhanced the multivalent binding to ConA. Consequently, the supramolecular structure of ligand–polyrotaxane conjugates yields the multivalent binding to proteins or receptors.

Cationic polymers are often used for gene delivery because of the Coulombic binding to anionic DNA and RNA. Various cationic cytocleavable polyrotaxanes have been developed by the modification of α- or β-cyclodextrin with cationic molecules such as oligoethyleneimine, *N,N*-dimethylaminoethyl group, and so on. The mobility and dissociation of cyclodextrin

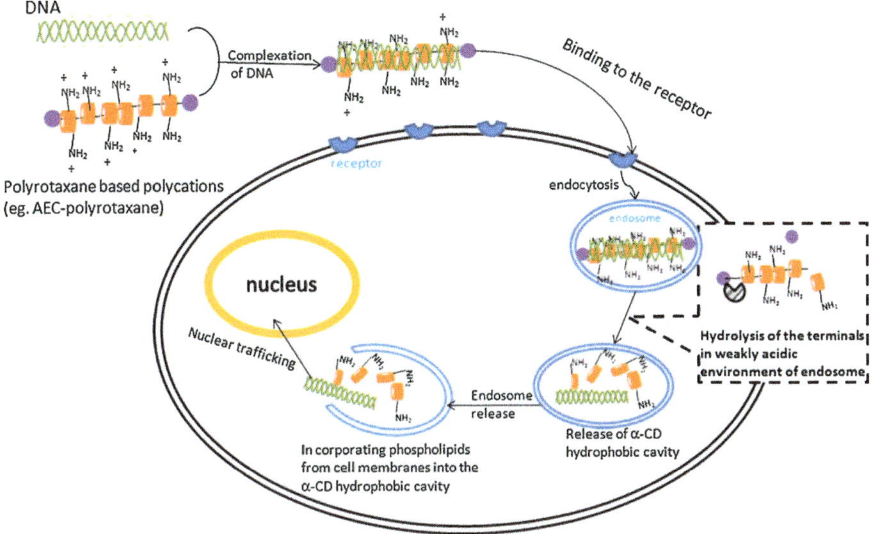

Figure 8.3   Strategic image of gene delivery using aminoethylcarbamoyl-(AEC)-polyrotaxane.[3]
Reproduced by permission of Springer.

are both effective for the strong complex formation of polyrotaxane and DNA, and to its ingenious intracellular DNA release, as shown in Figure 8.3.[8] Another cationic polyrotaxane was synthesized by grafting oligoethyleneimine on β-cyclodextrin threaded onto a pluronic poly(ethylene glycol)–poly(propylene glycol)–poly(ethylene glycol) triblock copolymer chain and capping the end with 2,4,6-trinitrobenzene sulfonate.[9] Such a block copolymer has some advantages: β-cyclodextrin selectively forms the inclusion complex with poly(propylene glycol) and not poly(ethylene glycol). Therefore, we can easily control the inclusion ratio of β-cyclodextrin in the polyrotaxane by changing the length ratio of poly(propylene glycol) to poly(ethylene glycol) in the Pluronic copolymer. The cationic polyrotaxane showed strong DNA binding ability and high gene transfection efficiency. The cationic cytocleavable polyrotaxane with N,N-dimethylaminoethylated α-cyclodextrins was also synthesized by introducing disulfide linkages between the backbone poly(ethylene glycol) and bulky end groups.[10] The *in vitro* dissociation experimental results demonstrated the sufficient cleavage of disulfide linkages, followed by rapid endosomal escape and plasmid DNA (pDNA) release.

Long poly(ethylene glycol) chains and α-cyclodextrin form pseudopolyrotaxane hydrogels *via* self-assembly in water, as shown in Figure 8.4.[11] Such a hydrogel shows the thixotropy and sol–gel thermal transition because it is a physical gel and the hexagonally packed tubular structure of α-cyclodextrins forms fragile cross-linking. In practice, Li *et al.* applied the unique properties of pseudopolyrotaxane hydrogels to injectable

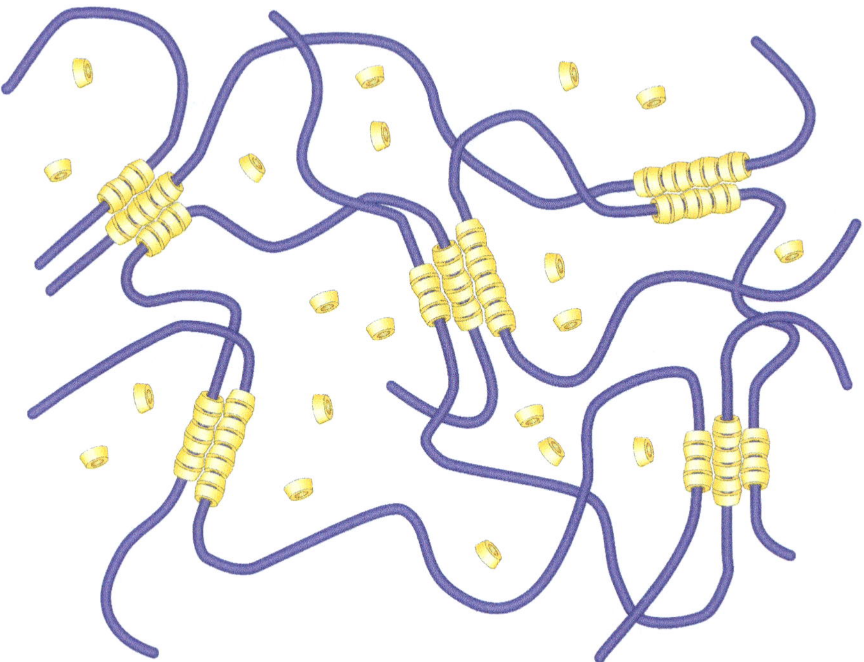

Figure 8.4   Supramolecular hydrogel formed by partial inclusion complexation of high molecular weight poly(ethylene glycol) and α-cyclodextrin.[3] Reproduced by permission of Springer.

drug delivery systems by incorporating some bioactive moieties, such as drugs, proteins, or plasmid DNAs, into the gel in a syringe.[3] Furthermore, the introduction of some biodegradable components, such as poly[(R)-3-hydroxybutylate], into the backbone copolymer is effective for the enhancement of the biosafety and biodegradability of the pseudopolyrotaxane hydrogel.[12] As a result, many cyclodextrin-based pseudopolyrotaxanes with different backbone copolymers have been synthesized so far.

Biodegradable cyclodextrin-based pseudopolyrotaxane/polyrotaxane hydrogels are also useful for tissue engineering because they can be gradually replaced by regenerated tissues. Yui and coworkers developed a variety of hydrolysable polyrotaxane hydrogels for tissue engineering.[13] For instance, the polyrotaxanes of α-cyclodextrin and poly(ethylene glycol) were cross-linked with other poly(ethylene glycol) chains.[13] The higher the cross-linking density, the slower the degradation of the polyrotaxane hydrogel. The polyrotaxane hydrogels showed larger cell adhesion and proliferation when they were cultured with fibroblasts. Moreover, highly porous polyrotaxane hydrogels were developed.[14] Since chondrocyte seeded in the porous polyrotaxane hydrogel scaffolds adhered and spread substantially in the pores, a cartilaginous extracellular matrix with a glycosaminoglycan fraction grew from the chondrocytes. The modification with carbonate-linked cholesterol on the porous polyrotaxane hydrogel scaffolds further enhanced the cell proliferation, glycosaminoglycan production, and the degradation of the porous polyrotaxane hydrogels.[15] And introducing primary amino groups in the porous polyrotaxane hydrogel scaffolds yielded a significant enhancement of chondrocyte attachment.[16] These results indicate that the biosafety and biodegradable pseudopolyrotaxane/polyrotaxane hydrogel is a powerful tool for tissue engineering.[3]

Thermo-responsive polymers have attracted much attention due to drastic change of the physical properties, such as the sol–gel transition, which is expected for biomedical devices, drug delivery systems, and so on. Some polyrotaxanes show thermos-reversible behavior at the molecular level: Harada and Kawaguchi reported a temperature-sensitive molecular shuttle in polyrotaxane, where a temperature change moved the cyclodextrin forward or backward along the polyrotaxane on different energetically stable sites.[17] Yui and coworkers also reported the thermo-reversible aggregation of α-cyclodextrin in a polyrotaxane with a backbone of Pluronic poly(ethylene glycol)– poly(propylene glycol)– poly(ethylene glycol) triblock copolymer.[18] At high temperature, α-cyclodextrins are localized in the poly(propylene glycol) component due to the hydrophobic interaction. This molecular change can be expanded to a sol–gel transition in the whole polyrotaxane solution.[19] It was reported that aqueous solutions of methylated polyrotaxane had a lower critical solution temperature (LCST) to form an elastic hydrogel as temperature increases. The X-ray diffractometry analysis indicated that the sol–gel transition of the methylated polyrotaxane solutions was due to formation and deformation of aggregates of methylated α-cyclodextrins in polyrotaxane, which corresponded to the hydrophobic dehydration and hydration, respectively. In the gel state, methylated α-cyclodextrins aggregate to form the cross-linking junction of the hexagonally packed channel structure, as shown in Figure 8.5. Clusters assembled with methylated α-cyclodextrins are gradually connected to the network with increasing temperature. The transition temperature strongly depends on the hydrophobic interaction strength of modified groups and on the modification ratio of the hydrophobic group on α-cyclodextrins. Therefore, the sol–gel transition temperature can be controlled in the thermo-responsive aqueous polyrotaxane solution. Another thermo-responsive polyrotaxane system was developed by cross-linking methacrylated polyrotaxane with *N*-isopropylacrylamide.[20]

8.2   Biosafety Test of Slide-ring Hydrogels

The slide-ring gel consisting of α-cyclodextrin and poly(ethylene glycol) end-capped with adamantane is expected to have a high biosafety and biocompatibility because all the component can be used for medical applications. Hence, the slide-ring gel is a promising candidate for wide applications *in vivo* as various devices or biomaterials. The method of examining the biocompatibility varies with the use. The US material test society (ASTM) defined more than 10 types of biocompatibility examinations depending on the purpose and the position of the test material.[21] Among the

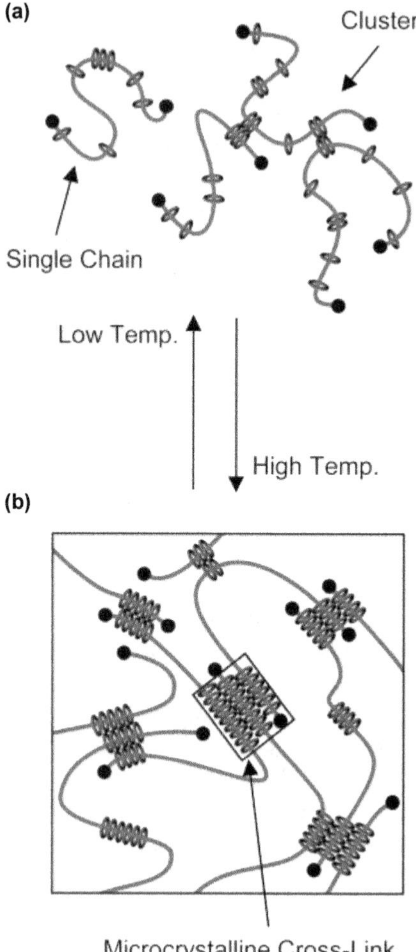

Figure 8.5   Schematic images of the sol–gel transition of the methylated polyrotaxane and water system. There are isolated chains and finite clusters of methylated polyrotaxane in the solution at lower temperature (a) and cross-links consisting of crystal-like aggregates among localized methylated α-cyclodextrins in the gel at higher temperatures (b).[19]
Reprinted with permission from *The Journal of Physical Chemistry B.* Copyright (2006) American Chemical Society.

examinations, biomaterials used for the contact lens or urethral catheter should be assessed by a stimulation study of the mucosa by direct contact of the test material and normal epithelium. In addition, Federation Dentaire International (FDI) established an implantation method for the detection of cytotoxicity by placing the test material in a subcutaneous place.[22] On the other hand, the Ministry of Health, Labour and Welfare, Japan gives "the guidelines of the fundamental biologic test" of medical devices and biomaterials. Several kinds of examinations for contact lenses are shown in the guideline, such as an inhibition test of the biomaterial regarding colony formation by extracted solutions, a maximization test of the animal *via* an extract or adjuvant patch test, a sub-acute toxicity test by the implantation method, a genetic toxicity examination, which is a back mutation examination, and a chromosome aberration examination. These examinations should be necessary for a new biomaterial to be used in practice. Advanced Softmaterials Inc. conducted implantation examination tests guided by FDI through various methods, described above, to find the most suitable for the living body adaptation test in the long-term.

To select the slide-ring gels that can be used for the long-term (from 3 months to 1 year), short-term (1 month) examination was performed for the first screening. Various forms of slide-ring gel were buried in the subcutaneous tissue on the backs of mice, and the buried material and surrounding tissue were examined histologically 1 month later. Then, long-term (3–6 months) examinations were conducted using the materials selected from the first screening.

To evaluate the compatibility of the slide-ring gel in the body, test materials of diameter 6 mm and thickness 1–2 mm were buried in the subcutaneous tissue on the backs of ICR mice for tissue analysis, as shown in Figure 8.6, and the abdomen for electron microscope observation, according to the guidelines of transplantation assays from FDI. The test materials of the fourth week were collected together with the surrounding tissue. The tissue section was made and analyzed by light microscopy. The compatibility of the materials was evaluated by the following points: infiltration of the inflammatory markers, measurement of the inflammation width, the antibody response, and the electron microscopy examinations. The slide-ring gels used in the biocompatibility test are summarized in Table 8.1.

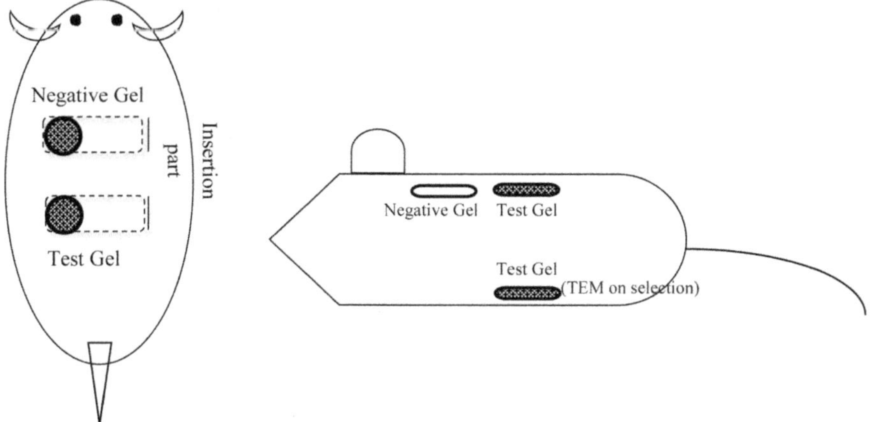

Figure 8.6   Safety and biocompatibility tests with slide-ring gels 6 mm in diameter and 1–2 mm in thickness, which were buried in the subcutaneous tissue on the backs of ICR mice (TEM, transmission electron microscopy).

Table 8.1 List of slide-ring gels used in the screening test. The slide-ring gels were prepared by gelation of a polyrotaxane (PR), methylated polyrotaxane (MPR), or hydroxypropylated polyrotaxane (HPR) with a cross-linker such as carbonyldiimidazole (CDI), divinyl sulfone (DVS), or butyl diglycol ether (BDGE) in dimethyl sulfoxide (DMSO) or sodium hydroxide aqueous solution. The solvent was replaced with low molecular weight poly(ethylene glycol) or water.

Number	Solvent[a]	PR[b]	Cross-linking method[c]	PR concentration on gelation	Cross-linker concentration	Final solid concentration
P1	PEG300	100-PR	CDI/DMSO	15%	1%	20%
P2	PEG300	35-MPR	CDI/DMSO	15%	1%	23%
P3	PEG300	35-MPR	DVS/NaOH	15%	1%	25%
P4	PEG300	35-MPR	BDGE/NaOH	15%	2%	15%
P5	PEG300	100-HPR	CDI/DMSO	15%	1%	14%
P6	PEG300	100-HPR	DVS/NaOH	15%	1%	17%
P7	PEG300	100-HPR	BDGE/NaOH	15%	2%	13%
W1	Water	35-MPR	CDI/DMSO	15%	1%	13%
W2	Water	100-HPR	DVS/NaOH	15%	1%	13%
W3	Water	35-PR	BDGE/NaOH	15%	3%	14%
W4	Water	50-HPR	BDGE/NaOH	15%	3%	11%

[a]PEG300, poly(ethylene glycol) with a molecular weight (M_w) of 300 g mol^{-1}.

[b]100-PR, polyrotaxane of α-cyclodextrin and poly(ethylene glycol) of M_w ca. 100 000 with a broad molecular weight distribution; 35-MPR, polyrotaxane of methylated α-cyclodextrin and poly(ethylene glycol) of M_w 35 000 with a narrow molecular weight distribution; 100-HPR, polyrotaxane of hydroxypropylated α-cyclodextrin and poly(ethylene glycol) of ca. M_w 100 000 with a broad molecular weight distribution; 35-PR, polyrotaxane of α-cyclodextrin and poly(ethylene glycol) of M_w 35 000 with a narrow molecular weight distribution; 50-HPR, polyrotaxane of hydroxypropylated α-cyclodextrin and poly(ethylene glycol) of ca. M_w 50 000 with a broad molecular weight distribution.

[c]CDI/DMSO, the polyrotaxane was cross-linked with carbonyldiimidazole as a cross-linker in dimethyl sulfoxide; DVS/NaOH, the polyrotaxane was cross-linked with divinyl sulfone as a cross-linker in sodium hydroxide aqueous solution; BDGE/NaOH, the polyrotaxane was cross-linked with butyl diglycol ether as a cross-linker in sodium hydroxide aqueous solution.

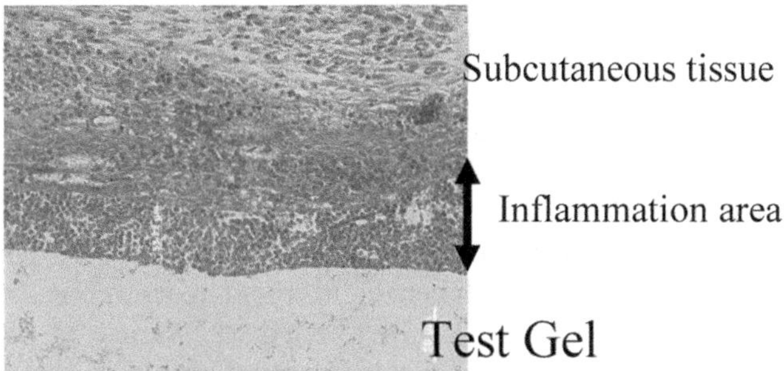

Figure 8.7   A typical TEM image of the conjunctive area between the test material and the surrounding tissue.

The inflammatory markers, such as lymphocytes and acidophilic leukocytes, were accumulated in the border area between the test material and the surrounding tissue. This accumulation width was measured as an inflammation index. The degree of the infiltration of inflammatory markers varied among the test materials. The test material using CDI as the cross-linking agent showed high infiltration of the inflammatory markers. DVS as the cross-linking agent exhibited less infiltration, and BDGE showed the least. A good correlation was seen between the composition of the buried material and the inflammatory response. Furthermore, the conjunctive area between the test material and the surrounding tissue was also analyzed by transmission electron microscopy (TEM), as shown in Figure 8.7. The test material with DVS as the cross-linker showed a weaker inflammatory response by TEM. This slide-ring gel showed no evidence of direct conjunction, such as the formation of a hemidesmosome, between the test material and the surrounding tissue, but erosion of the gel due to the invasion of macrophages was observed. Macrophages accumulated to remove inflammatory materials. For the long-term examination, the slide-ring gel with water or poly(ethylene glycol) as a solvent and BDGE as the cross-linker was selected because it was the least inflammatory.

Two and four weeks after implantation of the gel, the inflammatory area remained observable but accumulation of the phagocytes decreased after 1 month. At 8, 12, and 26 weeks, the inflammation area had disappeared. At 26 weeks, there were no deposits, and the border boundary surface between the gel and the tissue was smooth. Morphological analysis by transmission electron microscopy demonstrated that erosion and deposits due to macrophages were still seen 4 weeks after implantation, but deposits by macrophages diminished after 12–26 weeks. These results indicate the high biosafety and biocompatibility of slide-ring gels.

Another implantation test in a rabbit model showed that the slide-ring hydrogel was not adhered with even proliferated tissue, and thus seemed to have an inactive surface. In addition, no change was observed in the gel itself, which suggests that the native tissue did not invade into the gel. Such negligible interaction with tissues is one of the crucial features of slide-ring gels. As mentioned in Chapter 4, the movable cross-links in the slide-ring gel yield the peculiar mechanical properties such as the J-shaped stress–strain curve similar to mammalian skin, arteries, and so on. It shows extreme softness in the low elongation region, and stretch stiffening in the high elongation one. And the moduli in both regions can be controlled by changing the number of free rings and/or the cross-linking density in the slide-ring gel. Consequently, the slide-ring gel can be applied to a variety of biomaterials, including artificial arteries, artificial skin, intraocular lenses, and so on.

8.3   Other Applications of Slide-ring Gels

The slide-ring gel has movable cross-links, which are not only responsible for the pulley effect and peculiar mechanical properties but also for other physical properties. For instance, the ionic slide-ring gel shows a huge volume change up to 24 000 times by weight, and large overshoots on the swelling and shrinking processes.[23] Hence, the slide-ring gel is expected to show unique photo-responsive properties based on the movable cross-links. Such a photo-responsive slide-ring gel was actually developed by the modification of azobenzene on α-cyclodextrin in polyrotaxane.[23] By ultraviolet (UV) and visible light irradiation, azobenzene shows photo-isomerization between *cis* and *trans* isomers, which have different hydro-phobicities: the *trans* isomer induced by visible light is more hydrophobic than the *cis* one by UV. Accordingly, the light irradiation induces the volume change of azobenzene-containing hydrogels. The volume change behavior and the kinetics or dynamics in the swelling and shrinking processes in the photo-responsive slide-ring gel were compared to conventional photo-responsive chemical gels by cross-linking azobenzene-containing polymers.

Figure 8.8 shows the kinetics of the volume change of a photo-responsive slide-ring gel with three kinds of azobenzene derivatives in dimethyl sulf-oxide.[23] The gels have rapid swelling with a overshoot at an early stage within 60 min, followed by gradual shrinkage, which reached the steady state typically 1 day after light irradiation. The overshoot behavior at the transient stage is characteristic of slide-ring gels. This may be because the variable mesh size due to the movable cross-links enables the osmotic pressure to first drag excess amounts of solvent into the gel, followed by the subsequent sliding of cross-links to find the most favorable position, minimizing local tensions of network strands by the pulley effect, as mentioned in Chapter 4. The quite slow dynamics in the shrinkage process should reflect the latter process.

Alternate UV and visible light irradiation yields photo-induced expansion and contraction of the slide-ring gels in dimethyl sulfoxide, as shown in Figure 8.9.[23] Then, the maximum volume change reached 80–100%, which was larger than that of typical photo-responsive chemical gels, which range

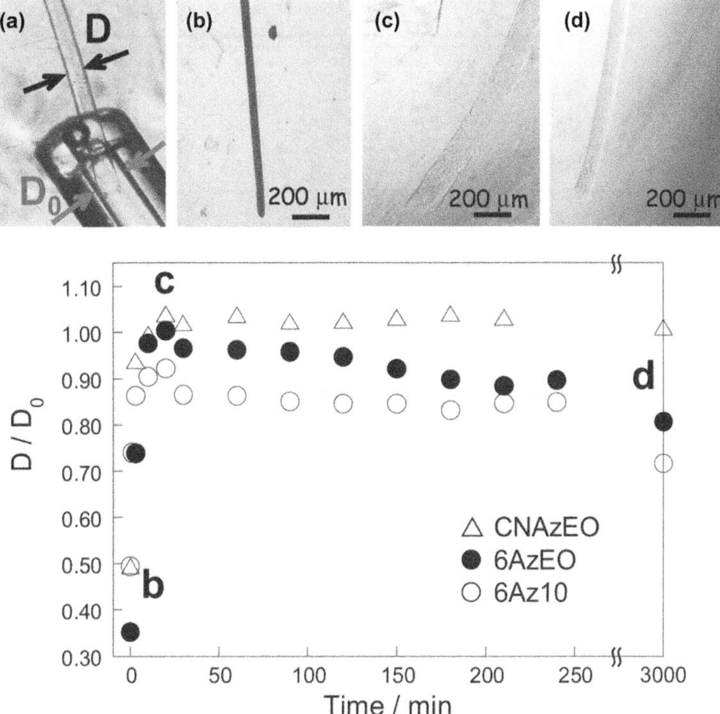

Figure 8.8 Swelling behavior of slide-ring gels in DMSO. (a) Preparation of the gel mold in a capillary cylinder. (b)–(d) Swelling behavior of azobenzen modified slide-ring gels in DMSO: initial dry state (b), intermediate overshooting state 20 min after immersion (c), and equilibrated state (d). The lower figure shows the time dependence of the swelling after immersion for three gels, where D is the diameter of the cylindrical gel and D_0 is that of the as-prepared one.[23]
Reproduced by permission of John Wiley & Sons, Inc.

from 10% to 30%. The large volume change should arise from the pulley effect based on the movable cross-links. Since chemically cross-linked gels have fixed junctions, the resulting microscopic inhomogeneity cannot be relaxed and suppresses full achievement of macroscopic deformation. When stimuli-responsive polymer gels have a periodical sub micron structure, they can be applied to active photonic band gap materials. A periodically porous slide-ring gel was also prepared by using a close-packed colloidal crystal as a template.[24] The chromic slide-ring gel exhibited solvatochromic behavior based on the change in structural color. The gel color changed from blue to red with decreasing water fraction of the solvent in the mixed solution of dimethyl sulfoxide and water.

The variable mesh size of the slide-ring gel is also expected to cause anomalies in the penetration flow properties through the gel membrane. The steady-state flow velocity of fluids through conventional porous membranes increases proportionally with imposed pressure, which is called

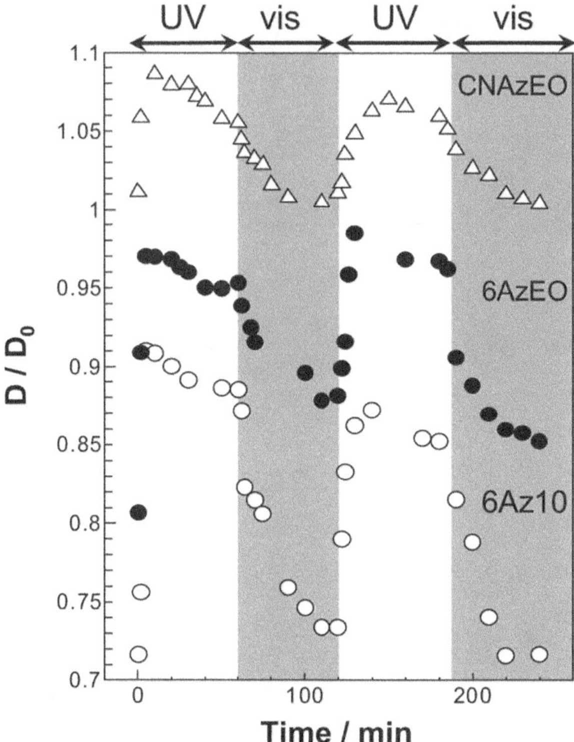

Figure 8.9   Swelling behavior of slide-ring gels in dimethyl sulfoxide upon alternate UV and visible light illumination.[23]
Reproduced by permission of John Wiley & Sons, Inc.

Darcy's law. It is well known that Darcy's law is satisfied in various chemical and physical gel membranes. The proportional constant reflects the friction coefficient between the gel network and fluid, which is usually independent of the induced pressure. This is because the mesh size of the chemical gel is not changed due to the fixed cross-linking. However, the slide-ring gel exhibited a peculiar dependence of the friction on the induced pressure, where the friction sharply changed between two different values within a narrow pressure range.[25]

Figure 8.10 shows the pressure dependence of the steady-state flow velocity of fluids through the slide-ring gel membrane. A transition region between two linear dependences is observed in both slide-ring gels with water and dimethyl sulfoxide (DMSO) as a solvent. This indicates that the friction drastically changes below and above the transition pressure threshold, which increases with cross-linking density. Figure 8.11 exhibits the dependence of the inverse of the product $f\Sigma$ of the friction (f) and the swelling ratio ($f\Sigma$) on the pressure gradient (p/d) for each slide-ring gel.[25] Two plateaus of the reduced friction $f\Sigma$ are clearly observed almost independently of the cross-linking density. This indicates that the dependence

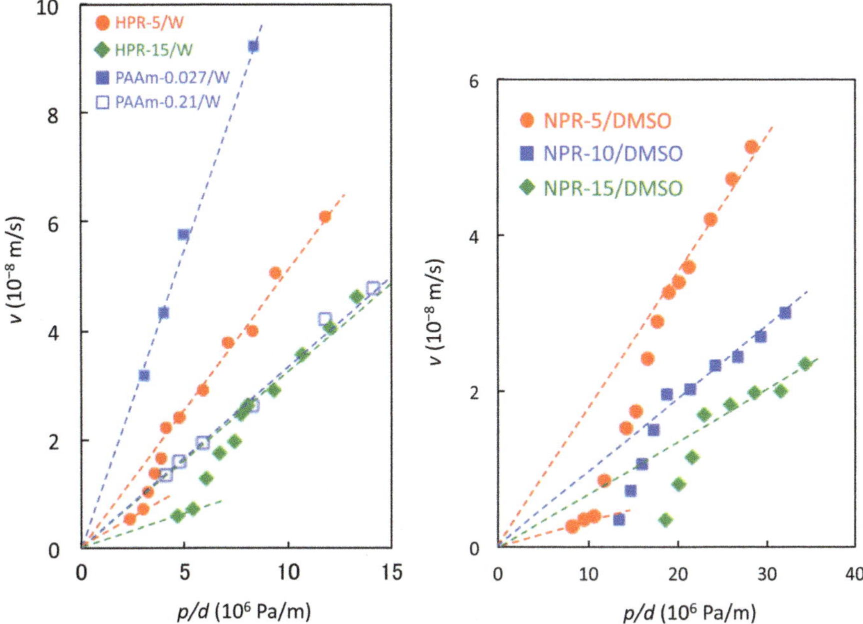

Figure 8.10 Pressure (p) dependence of the steady-state flow velocity (v) of fluids through the polymer gel membranes for hydroxylpropylated slide-ring gels (HPR) and polyacrylamide gel (PAAm) in water (W) (left), and for unmodified slide-ring gels (NPR) in dimethyl sulfoxide (DMSO) (right). The dashed lines depict the linear extrapolations passing through the origin (p/d, pressure gradient).[25]
Reproduced by permission of John Wiley & Sons, Inc.

of the plateau value on the cross-linking density mainly results from a difference in the fluid content of the gels. On the other hand, the difference between water and dimethyl sulfoxide in the plateau values can be explained by the solvent viscosity. Accordingly, the peculiar pressure-dependent permeation behavior of the slide-ring gel membrane is an inherent feature of the slide-ring gel with movable cross-links, which yield the variable mesh size.

The steep change of the friction in the narrow pressure range of the sliding gel suggests that the flow channel varies at a transition pressure threshold. This is ascribed to a pressure-induced structural change of the slide-ring gel network. The network structure of the slide-ring gel is homogeneous at low pressure due to the active sliding motion of the cross-links and free rings of the α-cyclodextrin along network strands to maximize the distribution entropy of the rings. This is the reason why the friction is so large at low pressure compared to usual chemical gels. The homogeneous network structure of the slide-ring gel was supported by small angle neutron and X-ray scatterings, as mentioned in Chapter 3. On the other hand, the network structure of the slide-ring gel becomes heterogeneous at high

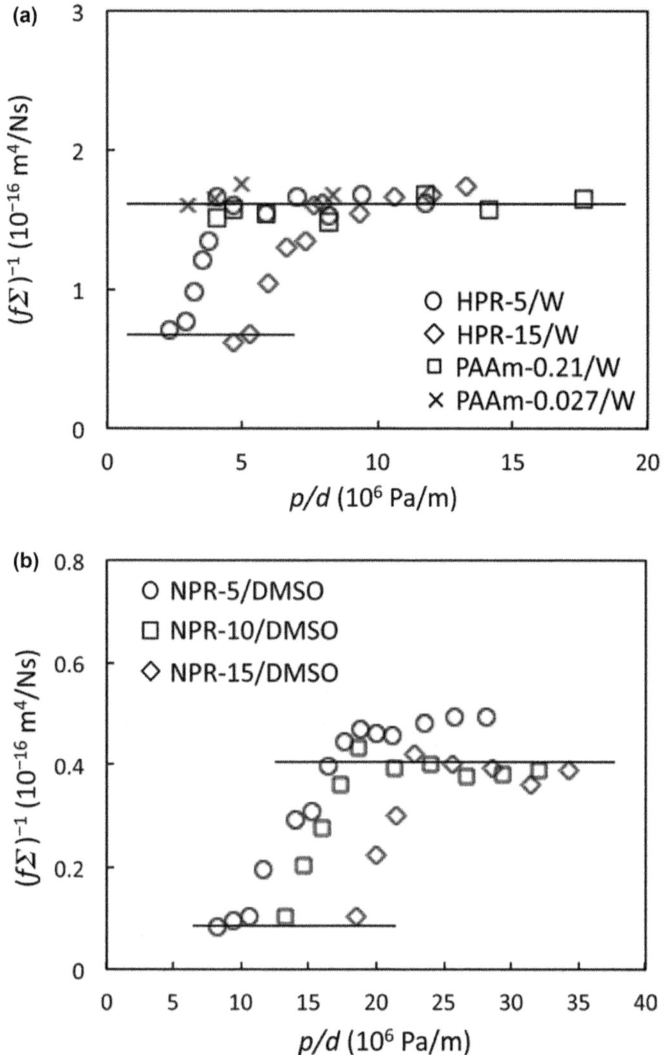

Figure 8.11 Inverse friction coefficient f multiplied by the degree of swelling Σ as a function of the pressure gradient p/d for HPR/W and PAAm/W gel membranes (a), and NPR/DMSO gel membranes (b). The product $f\Sigma$ for PAAm/W gels is independent of the cross-linker concentration. The same applies to $f\Sigma$ in the two plateau regimes for HPR/W and NPR/ DMSO gels.[25]
Reproduced by permission of John Wiley & Sons, Inc.

pressure, forming a larger flow channel with some aggregation of cross-links and free rings. Then, the slidability of the cross-links and free rings may be effectively suppressed by the distribution entropy. Consequently, the pressure-driven friction change in the slide-ring gel is ascribed to the flow-induced transition between a homogeneous and heterogeneous network

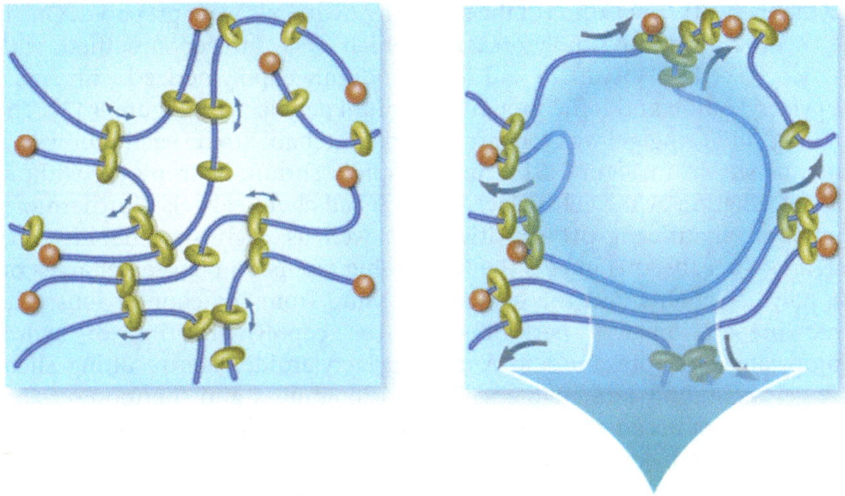

Figure 8.12 Schematic illustrations of the pressure-induced transition of the network structure of slide-ring gels, which have movable cross-links. At sufficiently low pressures, the network structure is homogenized by the active sliding motion of the cross-linked cyclodextrins thereby maximizing the arrangement entropy of the uncross-linked and cross-linked cyclodextrins. At pressures beyond the threshold, the network structure becomes heterogeneous due to the localization of cross-linked cyclodextrins with effectively suppressed slidability.[25]
Reproduced by permission of John Wiley & Sons, Inc.

structure accompanied by the slidability variation of cross-linked and free α-cyclodextrins (Figure 8.12).[25]

A sharp change in flow rate within a narrow pressure range for the slide-ring gels can be applied to a separation membrane and drug delivery system. The quasi on–off control of the permeation flow rate and controllable threshold is quite similar to the transistor in the electric circuit. Since the transistor was invented, the paradigm shift from analog to digital has occurred in the electric industry. The nonlinear transistor-like properties of the slide-ring gel may lead to a new drug delivery system that cannot be realized with usual chemical gels with only linear permeation flow properties. In addition, the variable mesh size and network structure in the slide-ring gel membrane can be extended to the peculiar transport properties of substances, including drugs in solution, as well as the nonlinear permeation behavior of fluids.

Many potential applications in multiple fields are expected in stimuli-responsive hydrogels changing their volumes and shapes, since these hydrogels are not so tough, although they are usually brittle. Takeoka *et al.* developed extremely stretchable thermo-sensitive hydrogels with good toughness by using polyrotaxane derivatives based on α-cyclodextrin and poly(ethylene glycol) as cross-linkers.[26] The has ionic groups, which promote the polyrotaxane cross-linkers to be well dispersed and extended in the polymer network. Two kinds of the slide-ring hydrogels were prepared. One is the hydroxypropylated polyrotaxane, which was further modified with 2-acryloyloxyethyl isocyanate and then randomly copolymerized with *N*-isopropylacrylamide and ionic monomer sodium acrylic acid (Figure 8.13). The obtained slide-ring gel was too strong and stretchable to be easily cut with a knife, as shown in Figure 8.14. And it rapidly shrank isotropically without any deformation at the gel surface unlike usual chemical gels. Furthermore, it showed significant pH sensitivity, as well as high stretchability and toughness because pH affects the Coulombic repulsion between charges on the polymer chains and the osmotic pressure from the counter ions. The other preparation was performed by the copolymerization of hydroxypropylated polyrotaxane and *N*-isopropylacrylamide. The resulting slide-ring gel was highly stretchable with good toughness and thermoreversible volume change because the cross-linked α-cyclodextrin can move along the backbone polyethylene glycol.

The polyrotaxane-based cross-linker developed indicates that the mechanical properties of polyelectrolyte hydrogels can be significantly improved without sacrificing their other characteristic properties. It is to be noted

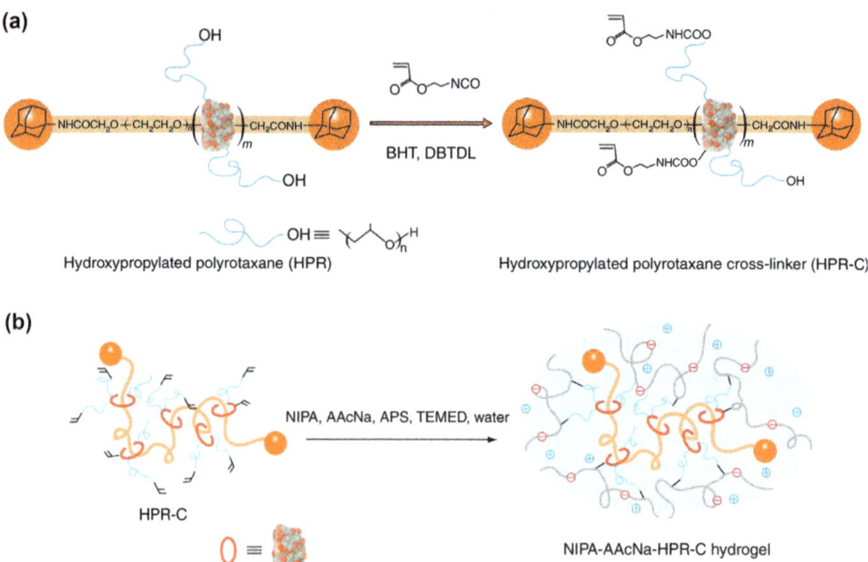

Figure 8.13   Preparation of the polyelectrolyte hydrogels using a nonionic polyrotaxane cross-linker. (a) Modification of hydroxypropylated polyrotaxane (HPR) with 2-acryloyloxyethyl isocyanate (HPR-C) (BHT, butylated hydroxytoluene; DBTDL, dibutyltin dilaurate). (b) Preparation of the slide-ring hydrogel (NIPA–AAcNa–HPR-C) from HPR-C (cross-linker) and *N*-isopropylacrylamide (NIPA) (main monomer) in water (APS, ammonium persulfate; TEMED, tetramethylethylenediamine).[26]
Reprinted by permission from Macmillan Publishers Ltd: *Nat. Commun.*, **5**, 5124, copyright (2014).

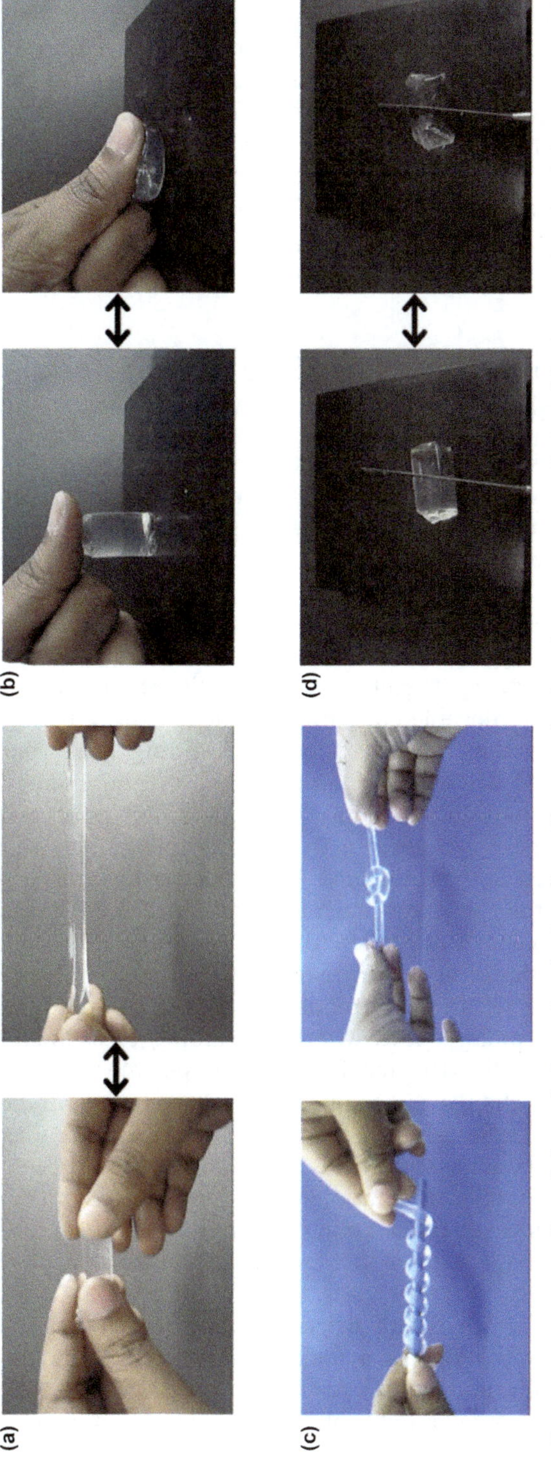

Figure 8.14 Properties of the polyelectrolyte hydrogels using a nonionic polyrotaxane cross-linker. (a) Elongated state of the NIPA–AAcNa–HPR-C hydrogel. (b) Compressed state of the NIPA–AAcNa–HPR-C hydrogel. (c) Coiled and knotted states of the NIPA–AAcNa–HPR-C hydrogel. (d) The NIPA–AAcNa–HPR-C hydrogel could not be easily cut with a knife.[26] Reprinted by permission from Macmillan Publishers Ltd: *Nat. Commun.*, 5, 5124, copyright (2014).

that the amount of polyrotaxane is much smaller than that of *N*-isopropylacrylamide. This means that we do not have to use only polyrotaxane to improve the mechanical strength of hydrogels from brittle to tough. A small amount of polyrotaxane as a cross-linker in the polymer network can change the mechanical properties drastically. And the tough slide-ring gels are simply synthesized with conventional polymer gel preparation techniques. Consequently, the concept of the movable cross-links can be applied to stimuli-sensitive hydrogels with good mechanical performance and controllable functionality for drug delivery systems, tissue engineering, artificial muscles, and sensors.

8.4   Other Applications of Slide-ring Materials in the Solid State

In the previous section, it turned out that a small amount of polyrotaxane as a cross-linker changed the mechanical properties of thermo-sensitive hydrogels remarkably. If this mechanical improvement strategy with polyrotaxane is effective for the solid state, the concept of the movable cross-links can be applied to elastomers, rubbers, and resins without solvent, as well as gels. As mentioned in Chapter 5, α-cyclodextrin in unmodified polyrotaxane forms aggregates in the solid state and cannot move freely along the backbone string. Therefore, some modification of polyrotaxane is required to apply the movable cross-links of the slide-ring materials to the solid state, *e.g.*, elastomers and resins. The sliding graft copolymer in Chapter 5 has some origomers on α-cyclodextrin, and thereby shows a melting temperature different from the unmodified polyrotaxane. This suggests that grafted α-cyclodextrin in polyrotaxane can move along the backbone string even in the solid state. When the sliding graft copolymer is cross-linked with other polymers, such as acrylate polymers, polyurethanes, and so on, the concept of movable cross-linking can be incorporated into the physical properties of usual polymeric materials, as well as the thermo-responsive slide-ring gels in the previous section. The movable cross-linking junctions yield the pulley effect, where the sliding cross-links along the backbone string equalize the tensions among polymer chains. And the movable cross-links change the chain length of a network strand between cross-links, so that uncross-linked free rings behave as air molecules in a one-dimensional tube and prevent the pulley effect to maximize the distribution entropy of the rings, as mentioned in Chapter 4. The pulley effect and entropy of the rings yields a low Young's modulus, small compression set, small stress relaxation, and large loss tangent in the viscoelastic profile in a wide frequency range.

The polyrotaxane grafted with origo-caprolactone can be soluble in a variety of organic solvents, has a melting temperature, and is currently produced industrially and commercially by Advanced Softmaterials Inc. Since origo-caprolactone-grafted polyrotaxane was commercialized, the applications of the slide-ring materials have been extended rapidly toward the solid state. For instance, they include scratch-resistant coatings, vibration- and sound-proof insulation materials for sound speakers, highly abrasive polishing media, dielectric actuators, pressure sensitive adhesives, shock resistance resins, and so on.[27]

In scratch-resistant coatings, origo-caprolactone-grafted polyrotaxane is cross-linked with suitable isocyanates and polyols, which can achieve a high cross-link density and low glass transition temperature at the same time. The high compatibility is quite important to formulate and realize excellent scratch resistance of the coated surface. The high elasticity and flexibility of the slide-ring materials (SRM) yield the shape memory effect that returns its original shape after deformation and repairs a scratch in the surface, as schematically shown in Figure 8.15. Small damage can thus be fixed immediately, as shown in Figure 8.16. It was found that, as well as the scratch resistance, the slide-ring material coating has strong adhesion to the substrates, which avoids crack and peel-off from the substrates. In addition, the scratch resistance of the slide-ring material coating was almost independent of temperature: it works well even at low temperature. The slide-ring material coating was originally developed as a top coat for automotive use, while it is now applied to mobile electronic devices, their accessories, and so on. Most recently, UV-curable slide-ring material coatings have been released to the market.

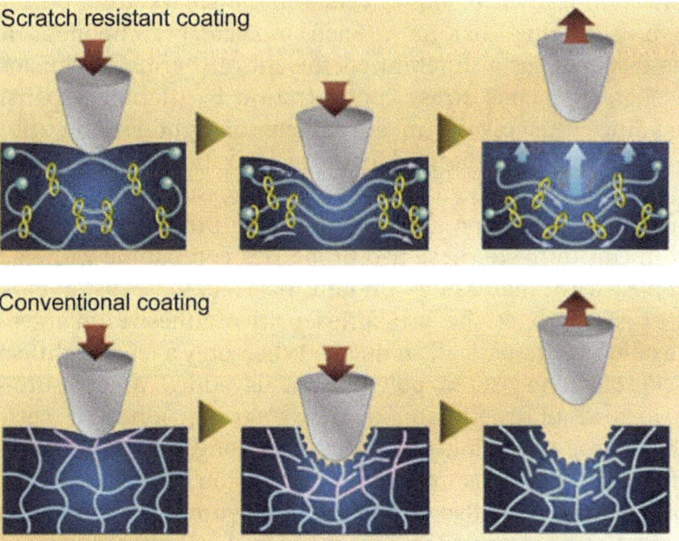

Figure 8.15   Schematic image of scratch recovery. Even with a high degree of cross-linking, the movable cross-links maintained the flexibility and elasticity of the coated layer, and this caused the scratch-healing properties.[27] Reproduced by permission of John Wiley & Sons, Inc.

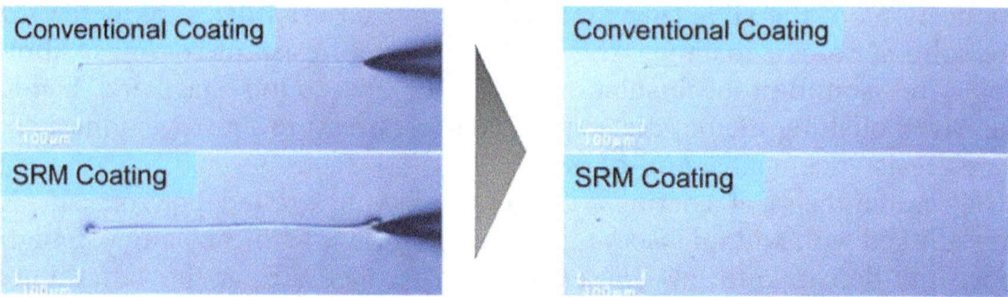

Figure 8.16   Self-recovery test of scratching. The self-recovery properties in a 15 μm thick SRM coating were confirmed by scratch testing with a sharp needle.[27]
Reproduced by permission of John Wiley & Sons, Inc.

The slide-ring material coating also shows a large loss tangent, especially in the high (over 8 kHz) frequency range, as shown in Figure 8.17. When it is used for a speaker cone, some unnecessary vibration of the cone should be reduced, which realizes clear and smooth sound. The slide-ring material coating for noise reduction is now on the market.[27]

Since the origo-caprolactone-grafted polyrotaxane is highly compatible with polyurethane, it can be easily incorporated into polymeric materials using polyurethane. Hence, a wide variety of applications have been developed in the fields of coatings, adhesives, sealants, and elastomers, as typical application areas of the polyurethane industry. For instance, pressure-sensitive adhesives for liquid crystalline displays are a promising application of slide-ring materials. Each molecular component in liquid crystalline displays has a different coefficient of thermal expansion, which yields the heterogeneous stress concentration by thermal expansion mismatching. Since this results in a problem of light leak in the display, pressure-sensitive adhesive is used for stress relaxation when liquid crystalline display panels are assembled. This kind of pressure-sensitive adhesive requires large stress relaxation and high adhesion properties at the same time but it is difficult to make these two properties compatible with each other. For example, with increasing amount of isocyanate as a cross-linker, the holding strength of the pressure-sensitive adhesive enlarges but the stress relaxation is dramatically reduced. When only a few weight percentage of origo-caprolactone-grafted polyrotaxane is added to pressure-sensitive adhesive instead of isocyanate, it shows strong adhesion performance without losing the relaxation properties and elongation.

In addition to the pressure-sensitive adhesion, the origo-caprolactone-grafted polyrotaxane can also be applied to polyurethane elastomers, which have extreme softness, quite low compression, low hysteresis loss, and high transparency. These properties are maintained in the slide-ring elastomers containing inorganic fillers or nanoparticles even at weight fractions of a few tens of percentage. As a result, a variety of unique items have been

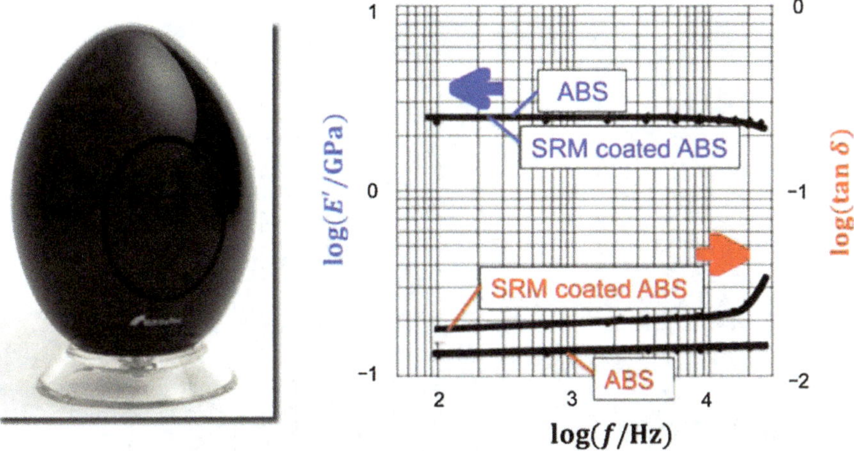

Figure 8.17   Egg-shaped speaker with a cone with an SRM-coated acrylonitrile–butadiene–styrene (ABS) resin. The right figure shows the difference between the cone with and without an SRM coating in the frequency dependence of the viscoelastic profiles. Although the SRM coating yielded no change in the storage modulus E', it increased the mechanical loss tangent (tanδ), especially in a higher frequency region.[27]
Reproduced by permission of John Wiley & Sons, Inc.

developed such as highly abrasive polishing media, dielectric actuators, and so on.

The slide-ring elastomer has been recently applied to highly abrasive polishing media, as shown in Figure 8.18.[27] The surface of the micrometer-scale slide-ring elastomer is coated with nano-size inorganic, hard abrasives. This cutting-edge polishing media is struck on the metal surface with a conventional blasting machine, known as shot peening, which achieves a superior mirror surface finish of the metal surface. The polishing media is retrieved to reuse many times. Typical polyurethane cannot realize such a high polishing performance. This is explained by the following mechanism: the slide-ring elastomer has extreme softness and strong recovery force on deformation compared to usual polyurethane elastomers. Hence, the nanoparticle-coated slide-ring elastomer sprayed from the processing machine at high speed is deformed largely on the metal surface at collision and slides along the metal surface, polishing the surface with the inorganic nanoparticles. As a result, the polishing media uses collision energy effectively to polish the work surface. It is to be noted that any abrasive of various hardness can be supported on the surface of the slide-ring elastomer. This means that the technology can be applied to a wide variety of items, including work pieces, from resins to carbides or ceramics. The mirror surface finish of the surface on complicated shapes has been performed mainly by craftsmen by hand. This new technology using a

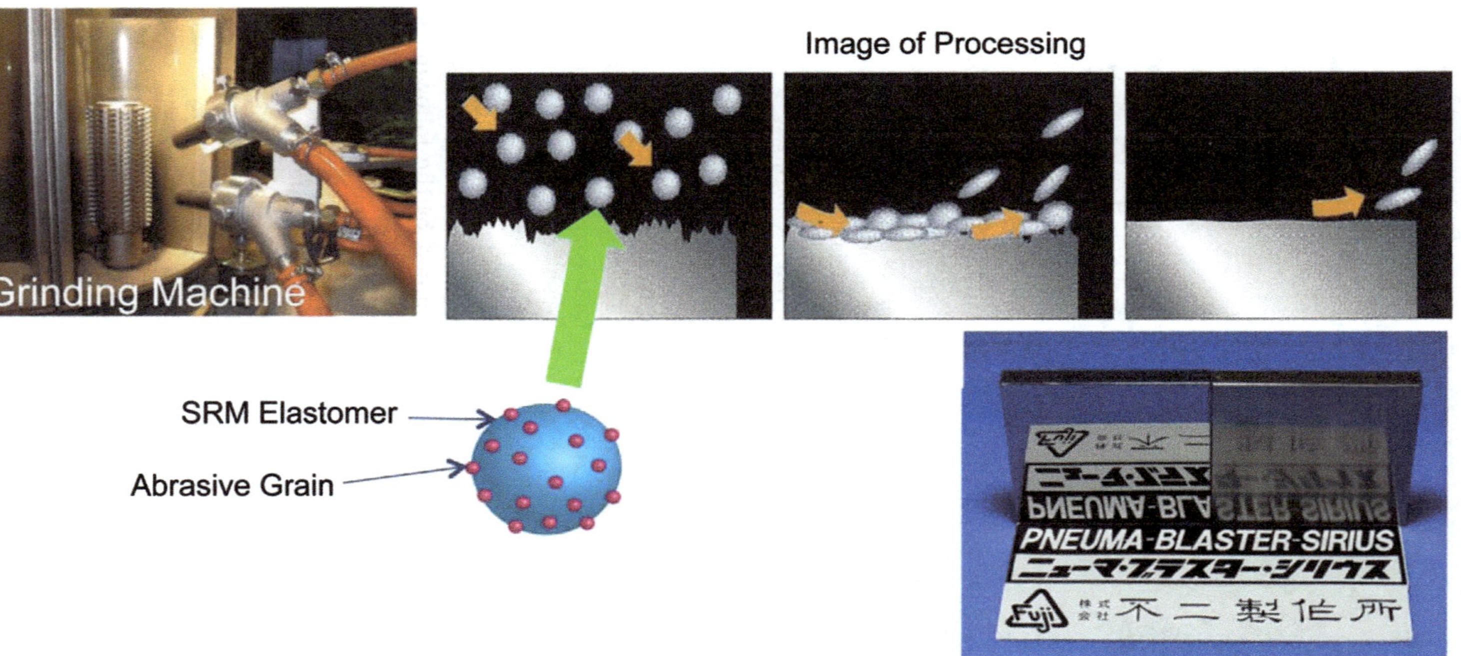

Figure 8.18 Highly abrasive polishing media with an abrasive supported on an SRM elastomer. The brand new polishing media could achieve a superior mirror surface finish with a processing machine that uses a kind of conventional blasting known as shot peening.[27]
Reproduced by permission of John Wiley & Sons, Inc.

slide-ring elastomer realizes automation of the mirror surface polishing on the complicated surface, which is a demand of many industries.

In recent years, many groups have proposed and developed various kinds of polymer actuators,[28] which are expected to be used for artificial muscle in robotics. Among them, dielectric polymer actuators consist of an insulating elastomer sandwiched by two flexible electrodes. The induced electric field squeezes the elastomer, of which the deformation is used for the actuator. The dielectric polymer actuator has many advantages, such as high power, lightness, large strain, quick response, high energy efficiency, and operation in air, compared to polymer gel actuators with solvent. Elastomers for the dielectric polymer actuator require physical properties of low modulus and high dielectric constant, which result in large strain under a low voltage, and good performance for the dielectric polymer actuator. The movable cross-links in the slide-ring material yield low modulus based on the J-shaped stress–strain curve, while the slide-ring elastomer has high dielectric constant of about 10 because of the high electric polarizability of α-cyclodextrin. Since the high dielectric constant is quite important for the dielectric polymer actuator, many dielectric polymer actuators are prepared by containing high dielectric constant fillers such as barium titanium oxide. Then, typical elastomers show a harder modulus and a larger hysteresis loop as the amount of filler increases. Hence, a low modulus is incompatible with a high dielectric constant in typical elastomers. However, the slide-ring elastomer exhibits less change in the modulus and hysteresis loop even with a high content of fillers.[29] Accordingly, the peculiar physical properties of the slide-ring elastomers have a great advantage over conventional rubbers or elastomers for dielectric polymer actuators. In addition, the slide-ring elastomer actuator can vibrate when an ac voltage is applied. The unique characteristic of the slide-ring elastomer can realize lighter, smaller, quieter, more sensitive, energy efficient, and high performance actuators, as shown in Figure 8.19.[27]

More recently, the applications of slide-ring materials have covered resins. Zhang *et al.* cross-linked an origo-caprolactone-grafted polyrotaxane, sliding graft copolymer (SGC), and polylactide with methylene diphenyl diisocyanate by reactive blending.[30] Polylactide (PLA) was transformed from a crystallized plastic into a totally amorphous elastomer as the cross-linking reaction with the origo-caprolactone-grafted polyrotaxane proceeded. The scanning and transmission electron microscopy analyses indicated that the elastic copolymer of polylactide and the origo-caprolactone-grafted polyrotaxane was formed at the interface to greatly improve the compatibility between the two domains of polylactide and origo-caprolactone-grafted polyrotaxane, and that the polylactide chains were extended. The resulting blends of polylactide, origo-caprolactone-grafted polyrotaxane, and methylene diphenyl diisocyanate (MDI) showed remarkable mechanical properties different from usual polylactide. Figure 8.20 shows the stress–strain curves of the blends of polylactide and

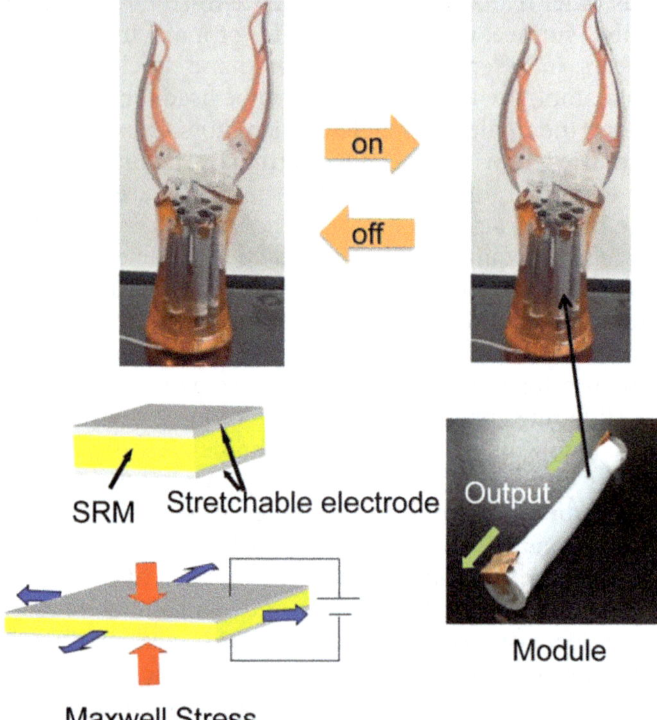

Figure 8.19   Dielectric polymer actuator with an SRM elastomer. An eight roll-type actuator inside the claw-shaped artificial hand stretched simultaneously as rapidly as the voltage was applied, and this resulted in the closing of the claws. The softness of the SRM elastomer realized a large strain under a low voltage.[27]
Reproduced by permission of John Wiley & Sons, Inc.

origo-caprolactone-grafted polyrotaxane with different amounts of methylene diphenyl diisocyanate.[29] As the amount of methylene diphenyl diisocyanate increases, the tensile toughness and strain at break enlarges dramatically. Figure 8.21 indicates the notched impact strength of these blends. Remarkably, the highest impact strength of the blend reaches about 20 times that of neat polylactide.[29] This means that the brittle polylactide was transformed to a tough resin by introducing origo-caprolactone-grafted polyrotaxane. This drastic improvement of the impact strength was attributed to the energy absorption and pulley effect of origo-caprolactone-grafted polyrotaxane, which cross-linked with polylactide, in the impact and deformation processes. This technology can also be applied to toughen other polymers, such as epoxy resin, polyester plastics, and polycarbonate, by adjusting the molecular structures of origo-caprolactone grafted-polyrotaxanes and interfacial agents.

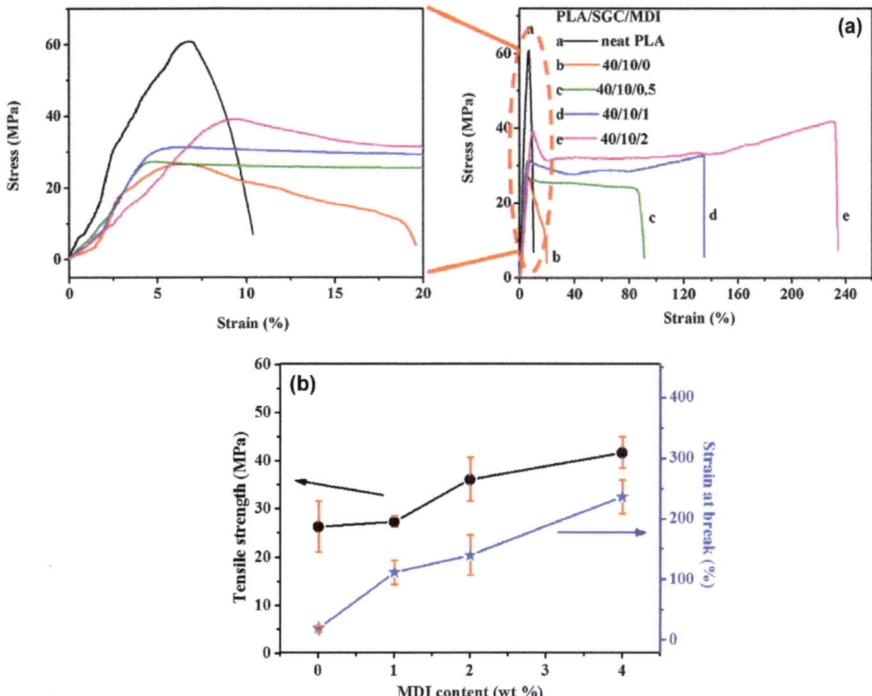

Figure 8.20 (a) Stress–strain curves, and (b) tensile toughness and strain at break of PLA/SGC and PLA/SGC/MDI with different MDI contents.[30]
Reprinted from *Polymer*, **55**, 4313, Copyright (2014), with permission from Elsevier.

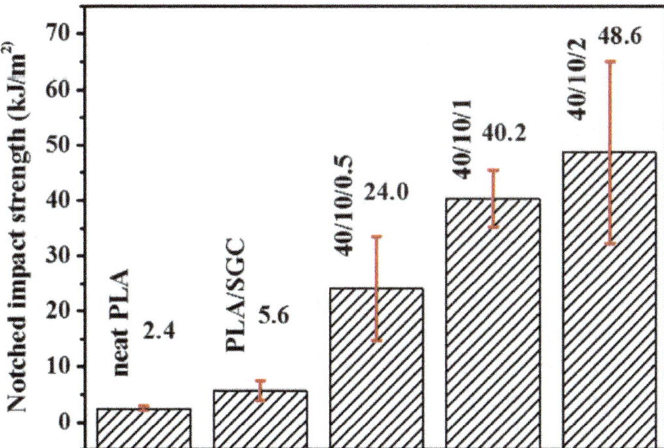

Figure 8.21 Notched impact strength of neat PLA, PLA/SGC, and PLA/SGC/MDI with different MDI contents.[30]
Reprinted from *Polymer*, **55**, 4313, Copyright (2014), with permission from Elsevier.

References

1. T. Ooya and N. Yui, *J. Biomater. Sci. Polym. Ed.*, 1997, **8**, 437; T. Ooya and N. Yui, *Macromol. Chem. Phys.*, 1998, **199**, 2311.
2. N. Yui and T. Ooya, *J. Artif. Organs.*, 2004, 7, 62.
3. J. J. Li, F. Zhao and J. Li, *Appl. Microbiol. Biotechnol.*, 2011, **90**, 427.
4. T. Ooya and N. Yui, *J. Controlled Release*, 1999, **58**, 251; T. Ooya and N. Yui, *Crit. Rev. Ther. Drug Carrier Syst.*, 1999, **16**, 289.
5. T. Ooya, T. Kawashima and N. Yui, *Biotechnol. Bioprocess Eng.*, 2001, **6**, 293; T. Ooya and N. Yui, *J. Controlled Release*, 2001, **80**, 219.
6. N. Yui, T. Ooya, T. Kawashima, Y. Sait, I. Tamai, Y. Sai and A. Tsuji, *Bioconjugate Chem.*, 2002, **13**, 582.
7. T. Ooya, M. Eguchi and N. Yui, *J. Am. Chem. Soc.*, 2003, **125**, 13016.
8. T. Ooya, A. Yamashita, M. Kurisawa, Y. Sugaya, A. Maruyama and N. Yui, *Sci. Technol. Adv. Mater.*, 2004, **5**, 363.
9. J. Li, C. Yang, H. Z. Li, X. Wang, S. H. Goh, J. L. Ding, D. Y. Wang and K. W. Leong, *Adv. Mater.*, 2006, **18**, 2969.
10. T. Ooya, H. S. Choi, A. Yamashita, N. Yui, Y. Sugaya, A. Kano, A. Maruyama, H. Akita, R. Ito, K. Kogure and H. Harashima, *J. Am. Chem. Soc.*, 2006, **128**, 3852.
11. J. Li, A. Harada and M. Kamachi, *Polym. J.*, 2006, **26**, 1019.
12. X. Li, J. Li and K. W. Leong, *Macromolecules*, 2003, **36**, 1209; X. Li, J. Li and K. W. Leong, *Polymer*, 2004, **45**, 6845.
13. J. Watanabe, T. Ooya, K. D. Park, Y. H. Kim and N. Yui, *J. Biomater. Sci., Polym. Ed.*, 2000, **11**, 1333; T. Ichi, J. Watanabe, T. Ooya and N. Yui, *Biomacromolecules*, 2001, **2**, 204.
14. T. Ichi, K. Nitta, W. K. Lee, T. Ooya and N. Yui, *J. Biomater. Sci., Polym. Ed.*, 2003, **14**, 567.
15. W. Tachaboonyakiat, T. Furubayashi, M. Katoh, T. Ooya and N. Yui, *J. Biomater. Sci., Polym. Ed.*, 2004, **15**, 1389.
16. T. Ooya, T. Ichi, T. Furubayashi, M. Katoh and N. Yui, *React. Funct. Polym.*, 2007, **67**, 1408.
17. Y. Kawaguchi and A. Harada, *Org. Lett.*, 2000, 2, 1353.
18. H. Fujita, T. Ooya, M. Kurisawa, H. Mori, M. Terano and N. Yui, *Macromol. Rapid Commun.*, 1996, **17**, 509; H. Fujita, T. Ooya and N. Yui, *Macromolecules*, 1999, **32**, 2534; T. Ikeda, N. Watabe, T. Ooya and N. Yui, *Macromol. Chem. Phys.*, 2001, **202**, 1338.
19. T. Kataoka, M. Kidowaki, C. Zhao, H. Minamikawa, T. Shimizu and K. Ito, *J. Phys. Chem. B*, 2006, **110**, 24377.
20. W. Li, Y. Guo, P. He, R. Yang, X. Li, Y. Chen, D. Liang, M. Kidowaki and K. Ito, *Polym. Chem.*, 2011, **2**, 1797.
21. ASTM standards, 1990.
22. 3 F-619 Standard Practice for Extraction of Medical Plastics.
23. T. Sakai, H. Murayama, S. Nagano, Y. Takeoka, M. Kidowaki, K. Ito and T. Seki, *Adv. Mater.*, 2007, **19**, 2023.
24. Y. Takeoka, H. Murayama, A. Imran, S. Nagano, T. Seki, M. Kidowaki and K. Ito, *Macromolecules*, 2008, **41**, 1808.

25. C. Katsuno, A. Konda, K. Urayama, T. Takigawa, M. Kidowaki and K. Ito, *Adv. Mater.*, 2013, **25**, 4636.
26. A. B. Imran, K. Esaki, H. Gotoh, T. Seki, K. Ito, Y. Sakai and Y. Takeoka, *Nat. Commun.*, 2014, **5**, 5124.
27. Y. Noda, Y. Hayashi and K. Ito, *J. Appl. Polym. Sci.*, 2014, **131**, 40509.
28. M. Kaneto, K. Kaneto, H. Fujisue, M. Kunifusa and W. Takashima, *Smart Mater. Struct.*, 2007, **16**, S250; E. Smela, O. Inganäs and I. Lundström, *Science*, 1995, **268**, 1735; Q. Pei and O. Inganäs, *Synth. Met.*, 1993, **55**, 3730; K. Asaka and K. Oguro, *J. Electroanal. Chem.*, 2000, **480**, 186; T. Fukushima, K. Asaka, A. Kosaka and T. Aida, *Angew. Chem. Int. Ed.*, 2005, **44**, 2410; R. Pelrine, R. Kornbluh, Q. B. Pei and J. Joseph, *Science*, 2000, **287**, 5454.
29. D. Yang, F. Ge, M. Tian, N. Ning, L. Zhang, C. Zhao, K. Ito, T. Nishi, H. Wang and Y. Luan, *J. Mater. Chem. A*, 2015, **3**, 9468.
30. X. Li, H. Kang, J. Shen, L. Zhang, T. Nishi, K. Ito, C. Zhao and P. Coates, *Polymer*, 2014, **55**, 4313.

Subject Index